Please
return to

Principles of
Dynamic Programming

CONTROL AND SYSTEMS THEORY

A Series of Monographs and Textbooks

Editor

JERRY M. MENDEL

University of Southern California
Los Angeles, California

Associate Editors

<table>
<tr><td>Karl J. Åström</td><td>Michael Athans</td></tr>
<tr><td>Lund Institute of Technology</td><td>Massachusetts Institute of Technology</td></tr>
<tr><td>Lund, Sweden</td><td>Cambridge, Massachusetts</td></tr>
</table>

David G. Luenberger
Stanford University
Stanford, California

Additional Volumes in Preparation

Principles of Dynamic Programming

Part I
Basic Analytic and Computational Methods

by ROBERT E. LARSON
Systems Control, Inc.
Palo Alto, California

JOHN L. CASTI
New York University
New York, New York

MARCEL DEKKER, INC.   New York and Basel

Library of Congress Cataloging in Publication Data

Larson, Robert Edward.
    Principles of dynamic programming.

    (Control and systems theory ; v. 7)
    Bibliography: v. 1, p.
    CONTENTS: pt. 1. Basic analytic and computational
methods.
    1. Dynamic programming. I. Casti, John, joint
author. II. Title.
T57.83.L37   519.7'03   78-15319
ISBN 0-8247-6589-3

MARCEL DEKKER, INC.

270 Madison Avenue, New York, New York 10016

Current printing (last digit):
10 9 8 7 6 5 4 3 2 1

PRINTED IN THE UNITED STATES OF AMERICA

To RICHARD BELLMAN

Teacher, Inspiration, and Friend to us both.

PREFACE

We are living in an age of ever-increasing complexity, insta-
bility, and uncertainty. Sociological, economic, and physical
pressures in all areas of modern life have generated an accelerated
demand for high-level decisionmaking based upon limited information
about the processes being controlled. In light of such inherent
uncertainties, decisionmaking procedures must remain flexible and
possess the capabilities for adapting to the changing needs of the
moment. In other words, regulation of a complex system is a multistage decisionmaking process carried out within the context of an
environment whose features are only partially known.

Several years ago, a systematic and concerted mathematical
study of such decisionmaking situations was initiated by Richard
Bellman. This pioneering work was based upon the fundamental system-
theoretic notion of feedback, i.e., that decision rules should be
based upon the current (and perhaps past) states of the process
under study. Living in the high intellectual culture of today,
this basic idea seems rather obvious; nevertheless, at the time it
represented a major conceptual advance, which enabled a decision-
maker to deal at one stroke with processes unfolding over time in
an uncertain manner. Bellman and his colleagues continued to
develop the feedback decisionmaking concept under the name of
"dynamic programming" and applied it to a wide variety of problems
in economics, engineering, operations research, and mathematics,
itself. It is probably true to say, though, that much of this
pioneering work was in advance of its time since the majority of

problems of true practical concern were computationally intractable
due to the limited state of the computing art at that time (circa
1960).

In recent years, a combination of rapid progress in computer
technology, coupled with the development of refined computational
procedures, has extended the range of problems amenable to a dynamic
programming treatment to such a degree that, for the first time,
truly significant problems can be attacked with confidence. Recent
crises in areas such as energy, the environment, industrial produc-
tivity, and economics have forced decisionmakers to examine their
options more carefully, and this examination has produced renewed
interest in utilizing methodological tools capable of providing
optimal policies for complex processes. Thus, the need for dynamic
programming is greater now than ever before.

The foregoing considerations make it clear that we are currently
at a most appropriate time for a reexamination of dynamic programming as a practical tool for solving significant problems in a
variety of fields. In response to this need, the current book
has been written to give an account of the basic analytic and
computational aspects of dynamic programming in a form accessible
to undergraduates. This book is intended as the first of two
volumes. The second volume will build on the material presented
here and will also deal with advanced topics and applications.

The current volume is devoted to the topic "Theory and Basic
Computational Procedures". The prerequisites for the material are
a semester course in ordinary differential equations and elementary
linear algebra. By keeping the background material to a minimum,
it is hoped that the book will be accessible to undergraduates,
possibly even in the sophomore year. Numerous solved problems are
also included to illustrate each key point made in the text. These
problems should be useful as models for more complex situations, as

well as providing deeper insight into the basic dynamic programming
procedures. Each chapter ends with a set of unsolved supplementary
problems which illustrate aspects of the theory not covered in the
earlier sections.

The current book is divided into four chapters. The first
chapter describes the basic properties of multistage decision
processes. The chapter begins with a review of the concept of a
differential equation, adds vector and matrix notation to the extent
needed in the volume, and develops the definition of a multistage
decision process in a step-by-step fashion.

The second chapter presents the basic theory of dynamic programming. In particular, the fundamental iterative equation of
dynamic programming is derived from basic principles. The significance of the equation is discussed in detail.

The third chapter discusses the basic computational procedure
of dynamic programming. All steps of the procedure are described
in detail, and a flow chart for a computer program to implement
the procedure is given. The steps are clarified through the detailed
working of an illustrative example. The computational implications
of the procedure are also examined.

In chapter four we treat various extensions of the basic procedures of the earlier chapters. Problems without an explicit
stage variable are discussed, and procedures for their solution are
developed. In addition, a number of auxiliary topics such as infinite stage processes, forward dynamic programming, and modified
computational procedures are also considered.

The second volume of the series will build on this basic
material to cover all aspects of dynamic programming. Volume II
describes the extension of the basic methods to problems where

uncertainty is present and indicates many application areas where
such effects are important. Both stochastic and adaptive methods
are developed. This volume also describes the numerous improvements
in the basic computational procedure that have been made over the
past 20 years. Other topics include the extension of dynamic programming to continuous-time systems, some fundamental analytic
results in the case of linear systems with quadratic criteria, and
some case studies of applications to actual large-scale, complex
problems.

The authors feel that these volumes will provide a comprehen-
sive treatment of all the major results obtained in the field of
dynamic programming. It is hoped that by compiling this material
and illustrating it with numerous examples, dynamic programming will
be brought to the attention of numerous researchers in a variety
of fields, assuring its central position in the family of problem-
solving approaches.

As in all undertakings of this type, heavy debts of gratitude
have been incurred through imposition on colleagues and students.
Much of the material has been tested, in one form or another, in a
course given over the past few years at Stanford University by one
of the authors (REL). Our grateful appreciation is extended to the
students in this course for acting as willing, although captive,
guinea pigs. In this same connection, we thank Professor S. Yakowitz
for his comments based upon feedback from a similar course given at
the University of Arizona. Finally, it is our great pleasure to
express gratitude to Professor Richard Bellman for his constant
encouragement and help during the course of this multi-year project.

PALO ALTO, CALIFORNIA   ROBERT E. LARSON
JULY 1977   JOHN L. CASTI

TABLE OF CONTENTS

Chapter 1

SYSTEMS, PROCESSES, AND DECISIONS

INTRODUCTION

The theory of dynamic programming may be considered the fundamental theory for the optimization of multistage decision processes.
Consequently, to lay the foundations for what follows, we begin our
exposition by giving a precise description of the concepts of a
multi-stage decision process. Through discussion of these fundamental topics, it will be seen that dynamic programming provides a
mathematical framework suitable for the consideration of far-reaching
generalizations of many classical problems, including significant
problems in the domains of engineering, physics, biology, economics,
and operations research, as will be amply illustrated in later
chapters.

VECTOR-MATRIX NOTATION

A subject of major importance in this book is the analysis of
systems described by several simultaneous differential equations.
In treating these systems it is convenient to make use of vector-
matrix notation. Let us now briefly summarize some of the definitions
and results that we shall require.

A column of n numbers

$$x = \begin{bmatrix} x_1 \\ x_2 \\ \vdots \\ x_n \end{bmatrix}$$

will be called an n-dimensional <u>vector</u>. The elements x_1, x_2,...,
x_n are called the components. Two vectors x and y will be considered equal when all of their respective components are equal.

   We define addition of two n-dimensional vectors as the vector
formed by adding componentwise, i.e.,

$$x+y = \begin{bmatrix} x_1 + y_1 \\ x_2 + y_2 \\ \vdots \\ x_n + y_n \end{bmatrix}$$

   Multiplication of a vector by a scalar (real or complex) α is
defined by the relation.

$$\alpha x = \begin{bmatrix} \alpha x_1 \\ \alpha x_2 \\ \vdots \\ \alpha x_n \end{bmatrix}$$

EXAMPLE

1.1 Define the vectors x and y as

$$x = \begin{bmatrix} 1 \\ 0 \\ -1 \end{bmatrix}, \qquad y = \begin{bmatrix} 2 \\ 3 \\ 0 \end{bmatrix}$$

Compute

(a)   x + y

(b)   2x

(c)   2x + 3y

(a)
$$x+y = \begin{bmatrix} 1 \\ 0 \\ -1 \end{bmatrix} + \begin{bmatrix} 2 \\ 3 \\ 0 \end{bmatrix} = \begin{bmatrix} 3 \\ 3 \\ -1 \end{bmatrix}$$

(b)
$$2x = 2 \begin{bmatrix} 1 \\ 0 \\ -1 \end{bmatrix} = \begin{bmatrix} 2 \\ 0 \\ -2 \end{bmatrix}$$

(c)
$$2x+3y = 2 \begin{bmatrix} 1 \\ 0 \\ -1 \end{bmatrix} +3 \begin{bmatrix} 2 \\ 3 \\ 0 \end{bmatrix} = \begin{bmatrix} 2 \\ 0 \\ -2 \end{bmatrix} + \begin{bmatrix} 6 \\ 9 \\ 0 \end{bmatrix} = \begin{bmatrix} 8 \\ 9 \\ -2 \end{bmatrix}$$

The transpose of a vector, denoted by a superscript T, is simply the quantities in the column array written as a row array. Thus, if a vector real x is defined as above, then

$$x^T = [x_1, x_2, \ldots, x_n]$$

Clearly, the transpose of the transpose of a vector is the vector itself.

The scalar product (or dot product) of two vectors is defined as the sum of the products of the components of the vectors. In other words, if x and y are two vectors of dimension n, then the scalar product, denoted as (x,y), takes the form

$$(x,y) = x_1 y_1 + x_2 y_2 + \ldots + x_n y_n$$

$$= \sum_{i=1}^{n} x_i y_i$$

Note from the definition that $(x,y) = (y,x)$.

EXAMPLE

1.2 Using the vectors x and y defined in Example 1.1 and defining

$$z = \begin{bmatrix} 0 \\ 0 \\ 1 \end{bmatrix} ,$$

compute

    (a) (x,y)

    (b) (y,z)

    (a) $(x,y) = 1 \cdot 2 + 0 \cdot 3 + (-1) \cdot 0 = 2$

    (b) $(y,z) = 2 \cdot 0 + 3 \cdot 0 + 0 \cdot 1 = 0$

A rectangular array of numbers having m rows and n columns

$$A = \begin{bmatrix} a_{11} & a_{12} & \cdots & a_{1n} \\ a_{21} & a_{22} & \cdots & a_{2n} \\ \vdots & & & \\ a_{m1} & a_{m2} & \cdots & a_{mn} \end{bmatrix} = \{a_{ij}\}; \quad i=1,2,\ldots m, \quad j=1,2,\ldots n$$

is called an mxn _matrix_. If m=n, A is termed a square matrix.
Addition of two mxn matrices is defined as

$$A + B = (a_{ij} + b_{ij}); \quad i=1,2,\ldots m, \quad j=1,2,\ldots,n$$

while scalar multiplication is given by

$$\alpha A = (\alpha a_{ij}); \quad i=1,2,\ldots,m, \quad j=1,2,\ldots,n$$

The _transpose of a matrix_, denoted by a superscript T, is the
nxm array obtained by exchanging the element in the i-th row, j-th
column with the element in the j-th row, i-th column. Formally, if
$A = \{a_{ij}\}$, then $A^T = \{a_{ji}\}$. Note that the definition for the transpose of a vector is just a special case of this equation.

Multiplication of an n-dimensional vector x by an mxn matrix A
is defined so that the set of m linear algebraic equations

$$\sum_{j=1}^{n} a_{ij} x_j = b_i, \quad i=1,2,\ldots,m$$

may be written as

$$Ax = b$$

Thus, Ax is defined to be the vector whose i-th component is

$$\sum_{j=1}^{n} a_{ij} x_j, \quad i=1,2,\ldots,m$$

Note that if y is an n-dimensional <u>row</u> vector, i.e., a 1xn matrix, then the product $y^T x$ is equal to

$$\sum_{i=1}^{n} y_i x_i$$

which is a scalar. Note also that if x and y are two column vectors, then the products $x^T y$ and $y^T x$ are equal to each other and to the scalar product (x,y) defined earlier.

Multiplication of two matrices is defined by considering the matrix Ax to be a linear transformation on x and examining the effects of two successive linear transformations on x. If A is an mxr matrix and B is an rxn matrix, then we define the matrix product AB to be an mxn matrix whose (i,j) component is

$$\{AB\}_{ij} = \sum_{k=1}^{r} a_{ik} b_{kj}, \quad i=1,2,\ldots m; \quad j=1,2,\ldots n$$

Multiplication is only defined for two matrices where the number of columns of the first is equal to the number of rows of the second.

EXAMPLE

1.3 Defining the vector $x = \begin{bmatrix} 1 \\ 0 \\ 1 \end{bmatrix}$ and the matrices $A = \begin{bmatrix} 1 & 0 & 0 \\ 2 & -1 & 3 \\ -4 & 1 & 2 \\ 0 & 0 & 1 \end{bmatrix}$

and $B = \begin{bmatrix} 2 & 0 & 1 \\ -1 & 0 & 3 \\ 0 & 1 & 0 \end{bmatrix}$

Compute

(a)   AB

(b)   Bx

(c)   A(Bx)

(d)   (AB)x

(a)

$$AB = \begin{bmatrix} 1 & 0 & 0 \\ 2 & -1 & 3 \\ -4 & 1 & 2 \\ 0 & 0 & 1 \end{bmatrix} \begin{bmatrix} 2 & 0 & 1 \\ -1 & 0 & 3 \\ 0 & 1 & 0 \end{bmatrix} = \begin{bmatrix} 2 & 0 & 1 \\ 5 & 3 & -1 \\ -9 & 2 & -1 \\ 0 & 1 & 0 \end{bmatrix}$$

(b)

$$Bx = \begin{bmatrix} 2 & 0 & 1 \\ -1 & 0 & 3 \\ 0 & 1 & 0 \end{bmatrix} \begin{bmatrix} 1 \\ 0 \\ -1 \end{bmatrix} = \begin{bmatrix} 1 \\ -4 \\ 0 \end{bmatrix}$$

(c)

$$A(Bx) = \begin{bmatrix} 1 & 0 & 0 \\ 2 & -1 & 3 \\ -4 & 1 & 2 \\ 0 & 0 & 1 \end{bmatrix} \begin{bmatrix} 1 \\ -4 \\ 0 \end{bmatrix} = \begin{bmatrix} 1 \\ 6 \\ -8 \\ 0 \end{bmatrix}$$

(d)

$$(AB)x = \begin{bmatrix} 2 & 0 & 1 \\ 5 & 3 & -1 \\ -9 & 2 & -1 \\ 0 & 1 & 0 \end{bmatrix} \begin{bmatrix} 1 \\ 0 \\ -1 \end{bmatrix} = \begin{bmatrix} 1 \\ 6 \\ -8 \\ 0 \end{bmatrix}$$

EXAMPLE

1.4   Define A and B as in Example 1.3 and matrix C as

$$C = \begin{bmatrix} 1 & 0 & 0 \\ 0 & 1 & 0 \\ 0 & 0 & 1 \end{bmatrix}$$

Compute

(a)   B A

(b) $B^T A^T$

(c) $(B^T A^T)^T$

(d) $B A^T$

(d)   A C

(e)   A C

(f)   C B

(a)   Since B is 3x3 matrix and A is a 4x3 matrix, the product is not defined.

(b)

$$B^T A^T = \begin{bmatrix} 2 & -1 & 0 \\ 0 & 0 & 1 \\ 1 & 3 & 0 \end{bmatrix} \begin{bmatrix} 1 & 2 & -4 & 0 \\ 0 & -1 & 1 & 0 \\ 0 & 3 & 2 & 1 \end{bmatrix}$$

$$= \begin{bmatrix} 2 & 5 & -9 & 0 \\ 0 & 3 & 2 & 1 \\ 1 & -1 & -1 & 0 \end{bmatrix}$$

(c)

$$(B^T A^T)^T = \begin{bmatrix} 2 & 0 & 1 \\ 5 & 3 & -1 \\ -9 & 2 & -1 \\ 0 & 1 & 0 \end{bmatrix}$$

Note from Example 1.4(a) that $AB = (B^T A^T)^T$. We will prove that this holds in general for any two matrices A and B in Solved Problem 1.5(a).

(d)
$$BA^T = \begin{bmatrix} 2 & 0 & 1 \\ -1 & 0 & 3 \\ 0 & 1 & 0 \end{bmatrix} \begin{bmatrix} 1 & 2 & -4 & 0 \\ 0 & -1 & 1 & 0 \\ 0 & 3 & 2 & 1 \end{bmatrix}$$

$$= \begin{bmatrix} 2 & 7 & -6 & 1 \\ -1 & 7 & 10 & 3 \\ 0 & -1 & 1 & 0 \end{bmatrix}$$

(e)
$$AC = \begin{bmatrix} 1 & 0 & 0 \\ 2 & -1 & 3 \\ -4 & 1 & 2 \\ 0 & 0 & 1 \end{bmatrix} \begin{bmatrix} 1 & 0 & 0 \\ 0 & 1 & 0 \\ 0 & 0 & 1 \end{bmatrix} \begin{bmatrix} 1 & 0 & 0 \\ 2 & -1 & 3 \\ -4 & 1 & 2 \\ 0 & 0 & 1 \end{bmatrix}$$

(f)
$$CB = \begin{bmatrix} 1 & 0 & 0 \\ 0 & 1 & 0 \\ 0 & 0 & 1 \end{bmatrix} \begin{bmatrix} 2 & 0 & 1 \\ -1 & 0 & 3 \\ 0 & 1 & 0 \end{bmatrix} \begin{bmatrix} 2 & 0 & 1 \\ -1 & 0 & 3 \\ 0 & 1 & 0 \end{bmatrix}$$

Note that AC = A and CB = B. A general result concerning matrices of the form of C is given in the next problem.

EXAMPLE

1.5 The nxn identity matrix I is defined to have value 0 off the diagonal and value 1 on the diagonal. ($I_{ij} = 0$, $i \neq j$; $I_{ii} = 1$, $i = 1,2,\ldots n$, $j=1,2,\ldots n$.) Prove that for any mxn matrix A, AI=A, and that for any nxm matrix B, IB=B.

Using the formula for matrix multiplication and the definition of I, we see that

$$\{AI\}_{ij} = \sum_{k=1}^{n} a_{ik} I_{kj}; \quad i=1,2,\ldots m, \quad j=1,2,\ldots n$$

$$= a_{ij}$$
$$= \{A\}_{ij}$$

Since AI and A are equal on a component-by-component basis, then we see that AI = A.

Similarly, we see that

$$\{IB\}_{ij} = \sum_{k=1}^{n} I_{ik} b_{kj}; \quad i=1,2,\ldots n, \quad j=1,2,\ldots m$$

$$= b_{ij}$$

$$= \{B_{ij}\}.$$

Again, since IB and B are equal on a component-by-component basis, then IB = B.

_____

If the components of a vector x or matrix A depend upon a variable t, we define differentiation and integration in the expected component-by-component fashion. Thus,

$$\frac{dx}{dt} = \begin{bmatrix} \dfrac{dx_1(t)}{dt} \\ \dfrac{dx_2(t)}{dt} \\ \cdot \\ \cdot \\ \cdot \\ \dfrac{dx_n(t)}{dt} \end{bmatrix}$$

$$\int x(t)dt = \begin{bmatrix} \int x_1(t)dt \\ \int x_2(t)dt \\ \cdot \\ \cdot \\ \cdot \\ \int x_n(t)dt \end{bmatrix}$$

EXAMPLE

1.6 Show that the solution to the matrix differential equation

$$\frac{dX}{dt} = AX, \quad X(0) = C$$

where A and C are constant $n \times n$ matrices and X is a time-varying $n \times n$ matrix, is given by

$$X(t) = e^{At}C,$$

where the matrix e^{At} is defined by the uniformly convergent series

$$e^{At} = \sum_{n=0}^{\infty} \frac{t^n}{n!} A^n$$

with

$$A^0 = I$$
$$A^1 = A$$
$$A^2 = A \cdot A$$
$$A^3 = A \cdot A \cdot A , \text{ etc.}$$

We show this directly by writing

$$\frac{dX}{dt} = \frac{d}{dt} \left(\sum_{n=0}^{\infty} \frac{t^n}{n!} A^n C \right)$$

$$= \frac{d}{dt} \left(I + \sum_{n=1}^{\infty} \frac{t^n}{n!} A^n \right) C$$

$$= \sum_{n=1}^{\infty} \frac{t^{n-1}}{(n-1)!} A^n \cdot C$$

$$= A \left(\sum_{n=0}^{\infty} \frac{t^n}{n!} A^n C \right)$$

$$= AX$$

$$X(0) = \sum_{n=0}^{\infty} \frac{0^n}{n!} A^n C$$

$$= IC$$

$$= C$$

EXAMPLE

1.7 Write the following n-th order linear differential equations
as systems of first-order differential equations of the form

$$\frac{dx}{dt} = Ax, \quad x(0) = c$$

where x and c are suitably-defined n-dimensional vectors and A
is an nxn matrix.

(a) $\dfrac{d^2x}{dt^2} + x = 0$, $x(0) = 0$, $\dfrac{dx}{dt}(0) = 1$

(b) $\dfrac{d^nx}{dt^n} + a_1\dfrac{d^{n-1}x}{dt^{n-1}} + a_2\dfrac{d^{n-2}x}{dt^{n-2}} + a_3\dfrac{d^{n-3}x}{dt^{n-3}} + \ldots$

$\quad + \ldots a_{n-1}\dfrac{dx}{dt} + a_n x = 0$

$\quad x(0) = c_1, \dfrac{dx}{dt}(0) = c_2, \ldots, \dfrac{d^{n-1}x}{dt^{n-1}}(0) = c_n$

(a) Since this is a second-order equation, we define x to be a
two dimensional vector

$$x = \begin{bmatrix} x_1 \\ x_2 \end{bmatrix}$$

where

$$x_1 = x$$
$$x_2 = \frac{dx}{dt}$$

Then, by definition,

$$\frac{dx_1}{dt} = x_2$$

From the second-order equation we have

$$\frac{dx_2}{dt} = \frac{d^2x}{dt^2} = -x = -x_1$$

In matrix-vector notation

$$\frac{d}{dt} \begin{bmatrix} x_1 \\ x_2 \end{bmatrix} = \begin{bmatrix} 0 & 1 \\ -1 & 0 \end{bmatrix} \begin{bmatrix} x_1 \\ x_2 \end{bmatrix}, \quad \begin{bmatrix} x_1(0) \\ x_2(0) \end{bmatrix} = \begin{bmatrix} 0 \\ 1 \end{bmatrix},$$

from which we identify

$$A = \begin{bmatrix} 0 & 1 \\ -1 & 0 \end{bmatrix}, \quad c = \begin{bmatrix} 0 \\ 1 \end{bmatrix}$$

(b) Defining

$$x = \begin{bmatrix} x_1 \\ x_2 \\ \vdots \\ x_n \end{bmatrix}$$

and

$$x_1 = x$$
$$x_2 = \frac{dx}{dt}$$
$$\vdots$$
$$x_n = \frac{d^{n-1}x}{dt^{n-1}}$$

we see that

$$\frac{d}{dt}\begin{bmatrix} x_1 \\ x_2 \\ x_3 \\ \vdots \\ x_{n-1} \\ x_n \end{bmatrix} = \begin{bmatrix} 0 & 1 & 0 & \cdots & 0 & 0 \\ 0 & 0 & 1 & \cdots & 0 & 0 \\ 0 & 0 & 0 & \cdots & 0 & 0 \\ \vdots & \vdots & \vdots & & & \\ 0 & 0 & 0 & \cdots & 0 & 1 \\ -a_n & -a_{n-1} & -a_{n-2} & \cdots & -a_2 & -a_1 \end{bmatrix} \begin{bmatrix} x_1 \\ x_2 \\ x_3 \\ \vdots \\ x_{n-1} \\ x_n \end{bmatrix}, \begin{bmatrix} x_1(0) \\ x_2(0) \\ x_3(0) \\ \vdots \\ x_{n-1}(0) \\ x_n(0) \end{bmatrix} = \begin{bmatrix} c_1 \\ c_2 \\ c_3 \\ \vdots \\ c_{n-1} \\ c_n \end{bmatrix}$$

from which the matrix A and the vector c can be deduced immediately.
Although this representation is not unique (any nonsingular coordinate transformation will produce an alternative representation),
it is a particularly convenient canonical form that we will use in
this book.

DYNAMICAL SYSTEMS

Let us now consider a particular type of multistage process
which has stimulated innumerable mathematical and engineering
investigations over the past two hundred years or so. This type
of process typically goes under the name of a dynamical system.

Consider the system of nonlinear ordinary differential
equations

$$\frac{dx}{dt} = f(x(t)), \quad x(0) = c \tag{1.1}$$

where x is a scalar and f is a function smooth enough to insure
that (1.1) possesses a unique solution for all $t > 0$. The
quantity c represents the initial condition of the system.

The basic idea we wish to employ in the study of dynamical
systems is to view the solution x not just as a function of t, but
also as a function of c, the initial condition; formally, $x = x(c,t)$.
In other words, we wish to study a particular solution of the
system (1.1) by imbedding it within a family of similar problems.

The foregoing idea leads us (via the assumed uniqueness of x) to study of the functional equation

$$x(c,\ s+t) = x(x(c,s),\ t). \tag{1.2}$$

Equation (1.2) is the mathematical statement of the fact that a system starting at c and evolving for a period of time s+t arrives at the same condition as a system which starts in $x(c,s)$ and evolves for a period of time t. This is a manifestation of the principle of causality, or semi-group property, for the solution of physical systems.

EXAMPLE

1.8 Consider the system

$$\frac{dx}{dt} = -x,\ x(0) = c.$$

Verify the causality principle for x.

Since the closed-form solution is

$$x(c,t) = e^{-t} c$$

we have

$$x(x(c,s),t) = e^{-t} (e^{-s} c)$$
$$= e^{-(s+t)} c$$
$$= x(c,s+t)$$

MULTISTAGE PROCESSES

Having obtained a feel for the consideration of the behavior of systems of differential equations as a function of both time and the specified initial state, let us examine a more general situation. Consider a point x in an abstract space or set X. For our present purposes, X will be a subset of n-dimensional Euclidean space R^n, but in more general situations this need not be the case. Let us call this subset the state space.

A particular point in this set X can be uniquely specified by its n coordinates in R^n. Let x denote the vector of these n coordinates.

$$x = \begin{bmatrix} x_1 \\ x_2 \\ \vdots \\ x_n \end{bmatrix}$$

We shall call the n components of x <u>state variables</u>, the vector x the <u>state vector</u>, and a specific value of x a <u>state</u>.

We now consider a transformation g of X into itself, which is applied to produce the sequence of states [x(0), x(1), x(2),...,] in the following manner

$$x(1) = g(x(0))$$
$$x(2) = g(x(1))$$
$$\vdots \qquad \vdots$$
$$x(k) = g(x(k-1)) = g^k(x(0))$$

The index of the sequence, i = 0, 1,..., k, will be called the <u>stage variable</u>.

Here we see that study of the behavior of the sequence [x(i)] is equivalent to a study of the successive iterates of the transformation g. The set of elements [x(0), x(1),...] will be termed a <u>multistage process</u>. Equivalently, we may denote the process by the pair (x(0),g), since these two items uniquely specify the sequence [x(i)].

EXAMPLE

1.9 Let x be the solution of the scalar differential equation

$$\frac{dx}{dt} = ax, \quad x(0) = c.$$

Let f be the transformation on R^1 defined by expressing $x(1)$ in terms of $x(0)$.

(a) What is the form of f?

(b) What is the behavior of f^n as $n \to \infty$?

(a) $x(t) = ce^{at}$, which implies that

$$x(1) = ce^a = e^a x(0).$$

Hence, the action of f is multiplication by e^a, i.e., $fb = e^a b$, $b \in R^1$.

(b) $f^n b = e^{na} b$

Thus, there are three possibilities for the limiting behavior of f^n as $n \to \infty$:

If $a > 0$, $\lim_{n \to \infty} f^n b = \infty$;

If $a < 0$, $\lim_{n \to \infty} f^n b = 0$;

If $a = 0$, $\lim_{n \to \infty} f^n b = b$.

Now let us examine a finite segment of the sequence $[x(i)]$. We agree to call the segment $[x(0), x(1),..., x(N)]$ an N-stage process. We wish to study the properties of different scalar functions of the elements $x(0)$, $x(1),...,x(N)$. For reasons that will become apparent in a later section, we shall call these functions criterion functions. To obtain results of any significance, we shall have to impose suitable structural assumptions on these functions since, as one might expect, little of consequence can come from consideration of a general abstract function.

Before proceeding to discuss these functions, let us mention that our previously discussed casuality relation (1.2) has a natural interpretation in the context of multistage processes. A

result of our assumptions on the mode of generation of the elements
[x(k)] is that given the state x(m), all subsequent elements are
uniquely determined by x(m) and f. In other words, "the future is
independent of the past", or, more prosaiclly, the future is determined only by the current state and the transformation law and not
by the particular path followed in reaching the current state.
Thus, as far as the state x(N) is concerned, it may be thought of
either as the Nth stage in an N-stage process starting in state
x(0) or the (N-m)th stage in an (N-m)-stage process which started
in state x(m). A moment's thought will convince the reader that
this is a generalized version of equation (1.2).

To initiate our study of the multistage process {x(0), g},let
us examine various properties associated with the particular criterion functions listed below:

$$\sum_{i=0}^{N} L(x(i)) \tag{1.3}$$

$$\prod_{i=0}^{N} L(x(i)) \tag{1.4}$$

$$L(x(N)) \tag{1.5}$$

$$\max_{0 \le i \le N} [L(x(i))] \tag{1.6}$$

In all cases L(x(i)) is a positive function of x(i). Let us make
use of the causality principle to derive recurrence relations
associated with the criterion functions (1.3) - (1.6). Consider
the function (1.3). At any intermediate state x, for any intermediate stage k, it is clear that the value of this criterion
can only be affected by the present and future states [x(k),
x(k+1),...x(n)]. Let us denote by S(x,k) that portion of the
criterion which can still be changed, where x(k) is the given
value x. Then we note that

$$S(x,k) = \sum_{i=k}^{N} L\,[x(k)] \tag{1.7}$$

Now consider the sum

$$\sum_{i=k+1}^{N} L[x(i)] = L[x(k+1)] + L[x(k+2)] +...+ L[x(N)]$$

$$= L[x(k+1)] + L[g(x(k+1)]+...+ L[g^{N-1}(x(k+1))] \tag{1.8}$$

By definition of the function $S(x,k)$, we see that the above sum is equal to

$$S(x(k+1),\ k+1) = S(g(x),\ k+1) \tag{1.9}$$

since $x(k+1) = g(x(k))$. Hence, we have the basic recurrence formula

$$S(x,k) = L(x) + S(g(x),\ k+1),\quad 0 \leq k < N \tag{1.10}$$

For k=N, we see that

$$S(x,N) = L(x) \tag{1.11}$$

EXAMPLE

1.10 Using the same chain of reasoning employed above, derive appropriate recurrence relations for the functions (1.4) - (1.6).

For the criterion $\prod_{i=0}^{N} L(x(i))$, the relations are

$$S(x,k) = L(x)S(g(x),k+1),\qquad 0 \leq k < N$$
$$S(x,N) = L(x)$$

For the criterion $L(x(N))$,

$$S(x,k) = S(g(x),\ k+1),\qquad 0 \leq k < N$$
$$S(x,N) = L(x)$$

For the criterion $\max_{0 \leq k \leq N}\ [L(x(k))]$,

$$S(x,k) = \max\ \{L(x),\ S(g(x),\ k+1)\},\quad 0 \leq k < N$$
$$S(x,N) = L(x)$$

So far we have always assumed that the transformation g was
the same each time it was applied. Suppose now we consider the
situation in which

$$x(k+1) = g(x(k),k) \qquad\qquad (1.12)$$

i.e., g is time-dependent. The reader will have little difficulty
convincing himself that the recurrence relations given above carry
over with only the change that g now becomes a function of time.
For example, (1.10) now becomes

$$S(x,k) = L(x) + S(g(x,k),k+1) \qquad\qquad (1.13)$$

while Eq. (1.11) remains unchanged. Also, no additional difficulty
occurs if the associated functions L are taken to be time-dependent,
i.e.,

$$L(x(k)) = L(x(k),k), \qquad\qquad (1.14)$$

In this case the relation (1.9) becomes

$$S(x,k) = L(x,k) + S[g(x,k), \; k+1] \qquad\qquad (1.15)$$

while Eq. (1.10) becomes

$$S(x,N) = L(x,N) \qquad\qquad (1.16)$$

MULTISTAGE DECISION PROCESSES

The stage is now set for us to consider a class of processes
which comprise a far-reaching generalization of those of classical
analysis. The theory of dynamic programming, which is the mathematical framework within which we study multistage decision processes, begins when we introduce the concept of decision-making into
our previous notions.

To introduce our ideas, let us begin with the discrete, deterministic multistage process $[x(0),g(x,k)]$ and assume that now the
transformation g depends not only upon the state x and the stage k,
but also upon a "decision" u, which is at our disposal, i.e.,
$g = g(x(k), u(k), k)$. We assume that the decision u is selected
from a set U which, for convenience, is assumed to be a subset of
m-dimensional Euclidean space R^m. We shall call the set U the
decision space. The coordinates of u can be arranged in a m-dimen-

sional vector called the decision vector.

$$
u = \begin{bmatrix} u_1 \\ u_2 \\ \vdots \\ u_m \end{bmatrix}
\tag{1.17}
$$

Again, the m components of u are called the decision variables, and
a specific value of u is called the decision.

Our multistage process now takes the form

$$x(1) = g(x(0), u(0), 0)$$
$$x(2) = g(x(1), u(1), 1)$$
$$\vdots$$
$$x(k+1) = g(x(k), u(k), k)$$
$$\vdots$$

(1.18)

The sequence of elements $\mathcal{U} = [u(0), u(1), \ldots, u(k), \ldots]$ is called
the decision sequence.

An immediate question that arises is the determination of a
rationale for selecting the decision sequence. Consequently, we
further postulate that there is given a scalar function J of the
states and decisions

$$J = J(x(0), x(1), \ldots, ; u(0), u(1), \ldots)$$

and that $\mathcal{U}$ is selected to minimize J. (More generally, the rationale could be based on some other evaluation of J, such as maximization.) The function J is called the cost function or criterion
function.

We note that an N-stage decision process is determined by
the elements

$$[x(0), x(1), \ldots, x(N); u(0), u(1), \ldots, u(N)]$$

where

$$x(k+1) = g(x(k),u(k),k), \quad k=0,1,\ldots N-1, \tag{1.19}$$

and where $x(0)$ is a given initial state. However, we see that all
$x(k)$, $k=1,\ldots,N$, can be expressed in terms of $x(0)$ and the $u(k)$,
$k=0,1,\ldots N-1$. Thus the criterion function could also be regarded
as having the form

$$J = J[x(0); u(0), u(1),\ldots u(N)]$$

If we follow the latter viewpoint and consider $x(0)$ to be a
given initial state and only the decisions $u(k)$, $k=0,1,\ldots N$, as free
variables, we could apply the necessary condition of calculus to
obtain the N simultaneous nonlinear equations

$$\frac{\partial J}{\partial u(k)} = 0, \quad i=0, 1,\ldots,N \tag{1.20}$$

Unless J is of a very simple nature, the resolution of (1.20) will
usually require sophisticated numerical search techniques. For
large N prohibitive computer times are the likely result of the
search procedures. Furthermore, rather severe requirements must be
placed upon J to ensure that the solution to (1.20) yields a
global minimum.

Thus, in terms of current analytic and computational capability,
it is extremely desirable to restrict our attention to classes of
multistage decision processes possessing certain structural features
which will allow us to determine the decisions $u(k)$ in a more
efficient manner. Fortunately, as we shall see, there are numerous
problems of great practical interest where we shall be able to use
the ideas of dynamic programming to solve for these decisions one
stage at a time.

We have seen that the most general type of multistage decision
process is outside our domain of approachable problems and that
restrictions on the form of the criterion function J are necessary
to make analytic and computational headway. One type of restriction
that may be made is to assume that the current decision $u(k)$ depends
only upon past states and decisions, i.e.,

$$u(k) = u(x(0),x(1),\ldots,x(k); \; u(0),\ldots,u(k-1); \; k) \qquad (1.21)$$

Such a function is called a _policy function_. A policy function giving decisions which minimize J is called an _optimal policy function_.

Unfortunately, there is still too much generality in (1.21) to be of practical value to us, and we will wish to concentrate our attention upon functions possessing the form

$$u(k) = u(x(k),k) \qquad (1.22)$$

That is, the current decision is to be a function only of the current stage and the current state. Needless to say, this is a very serious assumption, and its validity must be carefully established in any given situation.

We can see from the causality principle that the state of the system is indeed a complete summary of all past behavior. The assumption in Eq. (1.22) on the optimal policy function thus requires one further structural assumption on the criterion function J. The basic property J must possess to ensure that (1.22) holds is the _Markovian property_. This property means that after m decisions have been made, we wish the effect of the remaining N-m stages upon the total criterion function to depend only on the state x(m) and the final N-m decisions.

If we take the initial state x(0) as given and consider the criterion in the form

$$J = J \; [x(0), \; x(1),\ldots x(N); \; u(0), \; u(1),\ldots u(N)],$$

then a more rigorous statement of the Markovian property can be made as follows: we say that J has the Markovian property if, given the two decision sequences

$$\mathscr{U} = [u(0),u(1),\ldots,u(k),u(k+1),\ldots,u(N)]$$

and

$$\overline{\mathscr{U}} = [\bar{u}(0),\bar{u}(1),\ldots,\bar{u}(k),\bar{u}(k+1),\ldots,\bar{u}(N)]$$

then, whenever

$$x(k) = \bar{x}(k)$$
$$u(i) = \bar{u}(i), \quad i = 0, 1, \ldots, k-1$$

and

$$J(x(k), x(k+1), \ldots, x(N); \mathcal{U}) \leq J(x(k), \bar{x}(k+1), \ldots, \bar{x}(N); \mathcal{U}),$$

then

$$J(x(0), x(1), \ldots, x(N); \mathcal{U}) \leq J(x(0), \bar{x}(1), \ldots, \bar{x}(N); \overline{\mathcal{U}})$$

This statement of the Markovian property allows us to develop solutions of the type shown in Eq. (1.22) by using $x(k)$ to separate past decisions from current and future decisions. Note that the equality of $x(k)$ and $\bar{x}(k)$ and of $u(i)$ and $\bar{u}(i)$, $i=0,1,\ldots k-1$, together with the casuality of the state equations, imply that the past state sequences are equal i.e., that $x(i) = \bar{x}(i)$, $i=1,\ldots k$. Thus, the Markovian property is telling us not only that the current and future decisions $u(i)$, $i=k,\ldots N$, have no effect on past system behavior, but that their influence on the criterion function is independent of this past behavior, once the current state $x(k)$ is known.

An important issue to be resolved at this point is how often do the criterion functions obtained in actual practice possess the Markovian property. Fortunately, a large number of criterion functions that are of importance in engineering, economic, and biological control processes posses this vital property. In fact, we shall later see examples of criterion functions which cannot be expressed in analytic terms but which also possess the Markovian property.

As a first example of a class of criterion functions that have the Markovian property, let us consider the function

$$J = \sum_{i=0}^{N} L(x(i), u(i), i) \tag{1.23}$$

where again L is assumed to be a nonnegative function.

This function is called <u>separable</u> since it is the sum of terms
that depend only on the state and decision at a single stage.

EXAMPLE

1.11 Show that the separable criterion function

$$J = \sum_{i=0}^{N} L(x(i),u(i),i)$$

possess the Markovian property.

_____

Let $\mathcal{U}$ and $\overline{\mathcal{U}}$ be the two control sequences referred to above
and let

$$u(i) = \overline{u}(i) \text{ for } i=0,1,\ldots,k-1$$

where k is a fixed integer such that $0 \le k-1 \le N$. For J to have the
Markovian property, we must show that

$$x(k) = \overline{x}(k)$$

and

$$\sum_{i=k}^{N} L(x(i),u(i),i) \le \sum_{i=k}^{N} L(\overline{x}(i),\overline{u}(i),i)$$

implies

$$\sum_{i=0}^{N} L(x(i),u(i),i) \le \sum_{i=0}^{N} L(\overline{x}(i),\overline{u}(i),i)$$

To demonstrate this, write

$$\sum_{i=0}^{N} L(x(i),u(i),i) = \sum_{i=0}^{k-1} L(x(i),u(i),i) + \sum_{i=k}^{N} L(x(i),u(i),i)$$

$$= \sum_{i=0}^{k-1} L(\overline{x}(i),\overline{u}(i),i) + \sum_{i=k}^{N} L(x(i),u(i),i)$$

$$\leq \sum_{i=0}^{k-1} L(\overline{x}(i),\overline{u}(i),i) + \sum_{i=k}^{N} L(\overline{x}(i),\overline{u}(i),i)$$

$$= \sum_{i=0}^{N} L(\overline{x}(i),\overline{u}(i),i)$$

As a simple generalization of separability, consider the criterion function

$$J = \sum_{i=0}^{N} L(x(i+1), x(i), u(i), i). \qquad (1.24)$$

Since $x(i+1)$ is a function of $x(i)$ and $u(i)$ through the system dynamics

$$x(i+1) = g(x(i),u(i),i),$$

the function J is equivalent to a separable function.

As another generalization of separability, consider the function

$$J = \prod_{i=0}^{N} L(x(i), u(i), i) \qquad (1.25)$$

EXAMPLE

1.12   Show that the criterion function

$$J = \prod_{i=0}^{N} L(x(i),u(i),i)$$

possesses the Markovian property.   Recall that L is a positive function.

Let $\mathscr{U}$ and $\overline{\mathscr{U}}$ be the two control sequences referred to above and let

$$u(i) = \overline{u}(i) \text{ for } i = 0, 1, \ldots, k-1,$$

where k is a fixed integer such that $0 \leq k-1 \leq N$. For J to have the Markovian property, we must show that

$$x(k) = \bar{x}(k)$$

and

$$\prod_{i=k}^{N} L(x(i), u(i), i) \leq \prod_{i=k}^{N} L(\bar{x}(i), \bar{u}(i), i)$$

implies

$$\prod_{i=0}^{N} L(x(i), u(i), i) \leq \prod_{i=0}^{N} L(\bar{x}(i), \bar{u}(i), i).$$

As in Example 1.11, we write

$$\prod_{i=0}^{N} L(x(i), u(i), i)$$

$$= \left(\prod_{i=0}^{k-1} L(x(i), u(i), i) \right) \left(\prod_{i=k}^{N} L(x(i), u(i), i) \right)$$

$$= \left(\prod_{i=0}^{k-1} L(\bar{x}(i), \bar{u}(i), i) \right) \left(\prod_{i=k}^{N} L(x(i), u(i), i) \right)$$

$$\leq \left(\prod_{i=0}^{k-1} L(\bar{x}(i), \bar{u}(i), i) \right) \left(\prod_{i=k}^{N} L(\bar{x}(i), \bar{u}(i), i) \right)$$

$$= \left(\prod_{i=0}^{N} L(\bar{x}(i), \bar{u}(i), i) \right)$$

To dispel any notion that the Markovian property and separability are equivalent concepts, consider the criterion function

$$J = \sum_{i=0}^{k-1} |x(i)| + \left(\sum_{i=k}^{N} |x(i)|^2 \right)^{1/2}$$

This is a criterion function that possesses the Markovian property without being separable. Thus, although most of the criteria that one meets in practical problems are separable, our dynamic programming approach will be applicable to a broader class of problems.

SUMMARY

   At this point we have completely assembled the mathematical
framework we need in order to discuss the basic dynamic programming
procedure. In order to reach this degree of understanding, we
first discussed some basic properties of dynamical systems. We
then generalized these ideas to the concept of a multistage
process. Finally, we concluded the chapter by defining the multistage decision process and discussing a particular type of criterion
function, namely a function having the Markovian property, for
evaluating alternative decision sequences. In the next few chapters
we will build on this structure to develop the basic theory and
computational procedures of dynamic programming.

SOLVED PROBLEMS

1.1 Define the vectors x and y as

$$x = \begin{bmatrix} 1 \\ 0 \\ -1 \end{bmatrix} \qquad y = \begin{bmatrix} 2 \\ 3 \\ 0 \end{bmatrix}$$

(a) Compute (2x + 3y, x+y)

(b) Compute 2(x,x) + 5(x,y) + 3(y,y)

(c) Verify that (2x + 3y, x+y) = 2(x,x) + 5(x,y) + 3(y,y)

─────────────────────

(a) Compute

$$2x + 3y = 2 \begin{bmatrix} 1 \\ 0 \\ -1 \end{bmatrix} + 3 \begin{bmatrix} 2 \\ 3 \\ 0 \end{bmatrix} = \begin{bmatrix} 8 \\ 9 \\ -2 \end{bmatrix}$$

$$x + y = \begin{bmatrix} 1 \\ 0 \\ -1 \end{bmatrix} + \begin{bmatrix} 2 \\ 3 \\ 0 \end{bmatrix} = \begin{bmatrix} 3 \\ 3 \\ -1 \end{bmatrix}$$

Then,

    (2x + 3y, x+y) = 8·3 + 9·3 + (-2)(-1) = 53

(b) Compute

     (x,x) = (1)(-1) + 0·0 + (-1)(-1) = 2

     (x,y) = 2

     (y,y) = 13

Then 2(x,x) + 5(x,y) + 3(y,y) = 2·2 + 5·2 + 3·13 = 53

(c) By direct comparison

     (2x + 3y, x+y) = 53 = 2(x,x) + 5(x,y) + 3(y,y)

1.2 Define the matrices A, B and C as

$$A = \begin{bmatrix} 1 & 0 & 0 \\ 2 & -1 & 3 \\ -4 & 1 & 2 \\ 0 & 0 & 1 \end{bmatrix} \qquad B = \begin{bmatrix} 2 & 0 & 1 \\ -1 & 0 & 3 \\ 0 & 1 & 0 \end{bmatrix}$$

$$C = \begin{bmatrix} 1 & 0 & 0 \\ 0 & 1 & 0 \\ 0 & 0 & 1 \end{bmatrix}$$

Compute

(a) A + B

(b) B + C

(c) 2A

(d) A^T, B^T, C^T

(e) $2B^T + 3C$.

(a) Addition is not defined for two matrices having different dimensions. Since A is a 4x3 matrix and B is a 3x3 matrix, their sum is not defined.

(b)

$$B + C = \begin{bmatrix} 2 & 0 & 1 \\ -1 & 0 & 3 \\ 0 & 1 & 0 \end{bmatrix} + \begin{bmatrix} 1 & 0 & 0 \\ 0 & 1 & 0 \\ 0 & 0 & 1 \end{bmatrix}$$

$$= \begin{bmatrix} (2+1) & (0+0) & (1+0) \\ (-1+0) & (0+1) & (3+0) \\ (0+0) & (1+0) & (0+1) \end{bmatrix} = \begin{bmatrix} 3 & 0 & 1 \\ -1 & 1 & 3 \\ 0 & 1 & 1 \end{bmatrix}$$

(c)

$$2A = 2 \begin{bmatrix} 1 & 0 & 0 \\ 2 & -1 & 3 \\ -4 & 1 & 2 \\ 0 & 0 & 1 \end{bmatrix} = \begin{bmatrix} 2(1) & 2(0) & 2(0) \\ 2(2) & 2(-1) & 2(3) \\ 2(-4) & 2(1) & 2(2) \\ 2(0) & 2(0) & 2(1) \end{bmatrix}$$

$$= \begin{bmatrix} 2 & 0 & 0 \\ 4 & -2 & 6 \\ -8 & 2 & 4 \\ 0 & 0 & 2 \end{bmatrix}$$

(d)

$$A^T = \begin{bmatrix} 1 & 2 & -4 & 0 \\ 0 & -1 & 1 & 0 \\ 0 & 3 & 2 & 1 \end{bmatrix}$$

$$B^T = \begin{bmatrix} 2 & -1 & 0 \\ 0 & 0 & 1 \\ 1 & 3 & 0 \end{bmatrix}$$

$$C^T = \begin{bmatrix} 1 & 0 & 0 \\ 0 & 1 & 0 \\ 0 & 0 & 1 \end{bmatrix}$$

(e)

$$2B^T+3C = \begin{bmatrix} 7 & -2 & 0 \\ 0 & 3 & 2 \\ 2 & 6 & 3 \end{bmatrix}$$

1.3 Given that A, B, and C are all mxn matrices and that x and y are n-dimensional vectors, verify that

(a) (A+B) + C = A + (B+C)

(b) (A+B)x = Ax + Bx

(a) $(A+B)_{ij} = a_{ij} + b_{ij}$

$[(A+B) + C]_{ij} = a_{ij} + b_{ij} + c_{ij}$

$(B+C)_{ij} = b_{ij} + c_{ij}$

$[A + (B+C)]_{ij} = a_{ij} + b_{ij} + c_{ij}$

Since $[(A+B) + C]_{ij} = [A + (B+C)]_{ij}$ for all $i=1,2,\ldots,m$, $j=1,2,\ldots,n$, then

$(A+B) + C = A + (B+C).$

(b) $[(A+B)x]_i = \sum_{j=1}^{n} (a_{ij} + b_{ij}) x_j$

$[Ax + Bx]_i = \sum_{j=1}^{n} a_{ij} x_j + \sum_{j=1}^{n} b_{ij} x_j$

$= \sum_{j=1}^{n} (a_{ij} + b_{ij}) x_j$

Since $[(A + B) x]_i = [Ax + Bx]_i$ for all $i=1,2,\ldots,m$, then $(A + B)x = Ax + Bx.$

1.4   If A is an mxr matrix and B is an rxn matrix,
(a)   Prove that $(AB)^T = B^T A^T$.
(b)   Discuss under what conditions AB = BA.

─────────────────

(a)   Let C = AB, C an mxn matrix. Then $c_{ts} = \sum_{k=1}^{r} a_{tk} b_{ks}$, $t = 1,2,\ldots m$, $s = 1,2,\ldots n$.

Let $D = B^T A^T$, D an nxm matrix. Then

$d_{pq} = \sum_{k=1}^{r} [B^T]_{pk} [A^T]_{kq}$ $p = 1,2,\ldots n$, $q = 1,2,\ldots m$.

But

$$[B^T]_{pk} = b_{kp}, \cdot [A^T]_{kq} = a_{qk}$$

Therefore,

$$d_{pq} = \sum_{k=1}^{r} b_{kp} a_{qk} = \sum_{k=1}^{r} a_{qk} b_{kp} \quad q=1,2,\ldots m, \; p = 1,2,\ldots n.$$

Clearly, as q and r range from 1 to m and s and p range from 1 to n, whenever q = r = i and p = s = j, then

$$c_{ij} = d_{ij}$$

Therefore,

$$c^T = D$$

or

$$(AB)^T = B^T A^T$$

(b) If m ≠ n, then BA is not defined. If m = n, then BA has dimensions rxr, while AB is nxn. These matrices can be identical only if r = n.

If m = n = r, AB = BA if and only if

$$\sum_{k=1}^{n} a_{ik} b_{kj} = \sum_{k=1}^{n} b_{ik} a_{kj}, i=1,2,\ldots,n, \; j=1,2,\ldots n.$$

1.5 Equation (1.2) can be used to establish relations of elementary analysis, such as the trigonometric addition laws, exponential addition law, and the numerous functional equations satisfied by the so-called special functions, such as the recurrence relations for the Legendre, Hermite, and Laguerre polynomials, the addition formulas for the theta functions, and many others.

(a) Using the fact that the exponential function satisfies the equation

$$\frac{dx}{dt} = ax, \; x(0) = c$$

Show that

$$e^{a(s+t)}c = e^{at}(e^{as})c$$

(b)   Similarly, establish the addition law sin (t+s) = sin t cos s
+ sin s cos t by using the fact that sin t is the solution to the
equation

$$\frac{d^2x}{dt^2} + x = 0, \quad x(0) = 0, \quad \frac{dx}{dt}(0) = 1$$

(a)   From (1.2) we have

$$x(c, \; s+t) = e^{a(s+t)}c = x(x(c,s),t)$$
$$= e^{at}(e^{as}c)$$

(b)   Since sin (t+s) is a solution for any s, and since sin t and
cos t are fundamental solutions, we must have

$$\sin(t+s) = \alpha \sin t + \beta \cos t,$$

where α and β are functions of s.   Letting t = 0 gives β = sin s.
Differentiating and setting s = 0 then gives α = cos s.   Hence

$$\sin(t+s) = \cos s \sin t + \sin s \cos t.$$

1.6   Verify the causality principle,

$$x(c, s+t) = x(x(c,s), t),$$

for the differential equation

$$\frac{dx}{dt} = 1 + x^2, \quad x(0) = c.$$

HINT: The closed-form solution for x is $x(c,t) = \tan(t + \tan^{-1}c)$.

$$x(c,t) = \tan(t + \tan^{-1}c).$$

$$x(c, \; s+t) = \tan[s + t + \tan^{-1}c]$$

$$x(x(c,s),t) = \tan[t + \tan^{-1}[\tan(s + \tan^{-1}c)]]$$
$$= \tan[t + s + \tan^{-1}c].$$

Therefore,

$$x(c, \; s+t) = x(x(c, \; s), t)$$

1.7   Recall the functional equation

$$x(c, \ s+t) = x(x(c,s),t)$$

where x is the solution to

$$\frac{dx}{dt} = f(x(t)), \ x(0) = c$$

and x, c, and f are all n-dimensional vectors.   Under suitable
restrictions on the smoothness of x, establish that

$$\frac{\partial x}{\partial t} = \sum_{i=1}^{n} f_i(c) \frac{\partial x}{\partial c_i}$$

where f_i is the ith component of f.   Notice that this establishes
a relation between a _linear_ partial differential equation and the
original _nonlinear_ ordinary differential equations.   This illustrates a result due to Carlemann to the effect that the study of
any nonlinear system of equations may always be considered to be
equivalent to the study of an infinite set of linear equations.
The above formulation is often useful in ascertaining certain
structural features of the solution curve x, such as positivity,
monotonicity, and so forth.

For a small s, the Taylor series expansion yields

$$x(c,s) = x(c,0) + \frac{\partial x}{\partial t} (c,0)s + O(s^2)$$

$$= c + f(c)s + O(s^2)$$

$$x(c,s+t) = x(c,t) + s \frac{\partial x}{\partial t} (c,t) + O(s^2)$$

Thus, the functional equation (2.2) gives

$$x(c,t) + s \frac{\partial x}{\partial t} + O(s^2) = x(c+f(c)s + O(s^2),t)$$

$$= x(c,t) + s \sum_{i=1}^{N} f_i(c)\frac{\partial x}{\partial c_i} +O(s^2)$$

which, upon dividing through by s and letting s → 0, yields the
desired result.

1.8   Consider the linear differential equation $\dot{x}(t) + ax(t) = f(t)$, $x(0) = c$, a constant. Regard the solution as a transformation of t, x, and f, i.e., $x = h(t,c,f)$. Show that

$$h(t,c,f) = h_1(t,c) + h_2(t,f)$$

The solution to the differential equation has the form

$$x = ce^{-at} + \int_0^t e^{-a(t-t_1)} f(t_1) dt_1$$

Thus,

$$h_1(t,c) = ce^{-at} \text{ and}$$
$$h_2(t,f) = \int_0^t e^{-a(t-t_1)} f(t_1) dt_1$$

1.9   Consider the scalar transformation

$$x(k+1) = \alpha x(k) + b, \quad x(0) = c$$

Let

$$S(x,k) = \sum_{j=k}^{N} x^2(j), \text{ where } x(k) = x$$

(a)   Obtain a recurrence relation for $S(x,k)$
(b)   Show that $S(x,k)$ is a quadratic polynomial in x,
$$S(x,k) = r(k)x^2 + s(k)x + q(k)$$
(c)   Using the recurrence relation for $S(x,k)$, determine recurrence relations for $r(k)$, $s(k)$, and $q(k)$.
(d)   Determine the asymptotic behavior of the relations as $N \to \infty$.

(a)   We can use Eq. (1.10) and identify $L(x,k) = x^2$ and $g(x,k) = \alpha x + b$ to obtain

$$S(x,k) = x^2 + S(\alpha x + b, \ k+1)$$

(b)   Clearly the results hold if k=N.   Assume that S(x,k+1) is a quadratic polynomial in x, i.e.,

$$S(x,k+1) = r(k+1)x^2 + s(k+1)x + q(k+1)$$

Then,

$$
\begin{aligned}
S(x,k) &= x^2(k) + S[g(x,k),k+1] \\
&= x^2 + S[g(x,k), k+1] \\
&= x^2 + S(\alpha x + b, k+1) \\
&= x^2 + r(k+1)\,[\alpha x + b]^2 + s(k+1)\,[\alpha x + b] + q(k+1)
\end{aligned}
$$

Since this expression is clearly a quadratic polynomial in x, the proof is complete.

(c)   Substituting the quadratic form of S into the recurrence relation, we obtain

$$
\begin{aligned}
r(k)x^2 + s(k)x + q(k) &= x^2 + r(k+1)\,[\alpha x + b]^2 \\
&\quad + s(k+1)\,[\alpha x + b] + q(k+1) \\
&= x^2 + r(k+1)\,[\alpha^2 x^2 + 2\alpha x b + b^2] + s(k+1)\,[\alpha x + b] + q(k+1)
\end{aligned}
$$

Equating coefficients of like powers of c on both sides of this expression yields

$$
\begin{aligned}
r(k) &= 1 + \alpha^2 r\,(k+1) \\
s(k) &= 2\alpha b r(k+1) + \alpha s(k+1) \\
q(k) &= b^2 r(k+1) + b s(k+1) + q(k+1)
\end{aligned}
$$

Since

$$S(x,N) = r(N)x^2 + s(N)x + q(N) = x^2(N) = x^2$$

we have

$$
\begin{aligned}
r(N) &= 1 \\
s(N) &= 0 \\
q(N) &= 0
\end{aligned}
$$

(d)   A simple inductive argument easily establishes that

$$r(k) = \sum_{i=0}^{N-k} \alpha^{2i}$$

From this relation we see that for any finite k

$$\lim_{N\to\infty} r(k) = \begin{cases} \infty, & |\alpha|\geq 1 \\ \dfrac{1}{1-\alpha^2}, & |\alpha|<1 \end{cases}$$

Determination of the asymptotic behavior of s(k) and q(k) are left as optional exercises.

1.10 Develop versions of the equations in Example 1.10 on page 18 for time-dependent functions $g(x(k),k)$ and $L(x(k),k)$.

For the criterion $\displaystyle\prod_{k=0}^{N} L(x(k),k)$ with transformation $g(x(k),k)$ the relations are

$$S(x,k) = L(x,k) \; S(g(x,k), \; k+1), \; 0 \leq k < N.$$
$$S(x,N) = L(x,N).$$

For the criterion $L(x(N),N)$ with transformation $g(x(k),k)$, the relations are

$$S(x,k) = S(g(x,k), \; k+1), \; 0 \leq k < N$$
$$S(x,N) = L(x(N),N).$$

For the criterion $\displaystyle\max_{0\leq k\leq N} \; [L(x(k),k]$ with transformation $g(x(k),k)$, the relations are

$$S(x,k) = \max \; [L(x,k), \; S(g(x,k), \; k+1)], \; 0 \leq k < N$$
$$S(x,N) = L(x(N),N).$$

1.11 Show that the minimax criterion is Markovian. In this case, the objective is to minimize the maximum value that a particular function takes on during the decision interval. This type of problem is relevant to worst-case analyses of complex systems. Mathematically, the problem is to minimize

$$J = \max_{k=0,1,\ldots,N} \; [L(x(k), \; u(k), \; k)]$$

The problem can be stated as

$$\min_{u(0),u(1),..u(N)} J = \min_{u(0),u(1),...u(N)} \left\{ \max_{k=0,1,...,N} \left\{ L(x(k),u(k),k) \right\} \right\}$$

To establish the Markovian property, assume that

$$J(x(k),...,x(N); u(k),...u(N)) \leq J(\overline{x}(k),...,\overline{x}(N); \overline{u}(k),...\overline{u}(N))$$

$$x(k) = \overline{x}(k)$$

$$u(j) = \overline{u}(j) \quad \text{for } 0 \leq j \leq k-1$$

We then must prove that

$$J(x(0),...,x(N); u(0),...u(N)) \leq J(\overline{x}(0),...\overline{x}(N); \overline{u}(0),...\overline{u}(N))$$

The first assumption is equivalent to

$$\max_{k \leq j \leq N} \left\{ L(x(j),u(j),j) \right\} \leq \max_{k \leq j \leq N} \left\{ L(\overline{x}(j),\overline{u}(j),u) \right\}$$

The other two equations and the fixed initial state $x(0)$ imply that

$$x(j) = \overline{x}(j) \quad \text{for } 0 \leq j \leq k-1.$$

Therefore,

$$\max_{0 \leq j \leq k-1} \left\{ L(x(j),\ u(j),\ j) \right\} = \max_{0 \leq j \leq k-1} \left\{ L(\overline{x}(j),\overline{u}(j),j) \right\} = A$$

To prove our assertion, we see that

$$J(x(0),...x(N); u(0),...u(N)) = \max_{0 \leq j \leq N} \left\{ L(x(j),u(j),j) \right\}$$

$$= \max \left\{ \max_{0 \leq j \leq k-1} \left\{ L(x(j),u(j),j) \right\},\ \max_{k \leq j \leq N} \left\{ L(x(j),u(j),j) \right\} \right\}$$

$$= \max \left\{ A,\ J(x(k),...,x(N); u(k),...u(N)) \right\}$$

Similarly, we see that

$$J(\overline{x}(0),\ldots\overline{x}(N); \overline{u}(0),\ldots\overline{u}(N))$$

$$= \max \left\{A, \; J(\overline{x}(k),\ldots\overline{x}(N),\ldots\overline{u}(k),\ldots\overline{u}(N))\right\}$$

We can now use our first assumption to see that

$$J [x(0),\ldots x(N); u(0),\ldots u(N)]$$

$$= \max J \left\{A, \; J[x(k),\ldots x(N); u(k),\ldots u(N)]\right\}$$
$$\leq \max \left\{A, \; J[\overline{x}(k),\ldots\overline{x}(N); \overline{u}(k),\ldots u(N)\right\}$$
$$\leq J [\overline{x}(0),\ldots\overline{x}(N); \overline{u}(0),\ldots u(N)]$$

1.12 Show that the criterion function

$$J = x(0) \; u(0) \; u(1) + u^2(1)$$

does not have the Markovian property. HINT: Try to construct a counterexample to the Markovian property using the system equation $x(k+1) = x(k) + u(k)$, the state $x(1) = 0$, and decision sequences with $u(1) = 0$ and $\overline{u}(1) = b$, a positive constant.

We need to find two decision sequences, $\mathcal{U} = [u(0), u(1)]$ and $\overline{\mathcal{U}} = [\overline{u}(0), \overline{u}(1)]$, and a state $x(1)$, such that

$$J[x(1), u(1)] < J [x(1), \overline{u}(1)]$$

results in

$$J[x(0), x(1); u(0), u(1)] > J [x(0), x(1); u(0), \overline{u}(1)]$$

Following the hint, let us use the system equation $x(k+1) = x(k) + u(k)$, the state $x(1) = 0$, and the decision sequences $\mathcal{U} = [u(0), 0]$, $\overline{\mathcal{U}} = [u(0), b]$. We see immediately

$$J[0; u(1)] = 0 < b^2 = J[0; \overline{u}(1)]$$

However, from the system equation and the condition $x(1) = 0$, we see that $u(0) = -x(0)$.

Then

$$J[x(0),0; u(0),0] = 0$$

$$J[x(0),0; u(0),b] = -b\ x^2(0) + b^2$$

Clearly for any $x(0) \neq 0$, we can always choose a positive value b
such that

$$0 < b < x^2(0).$$

In this case

$$J[x(0),\ 0; \mathcal{U}] = 0 > -bx^2(0) + b^2 = J[x(0),\ 0; \overline{\mathcal{U}}]$$

Thus we have a violation of the basic requirement for the Markovian
property, implying this criterion function is not Markovian.

SUPPLEMENTARY PROBLEMS

1.13 Verify the causality principle for x, where x satisfies

$$\frac{dx}{dt} = 1 - x^2, \; x(0) = c$$

HINT: The closed-form solution for x is

$$x(c,t) = \tanh (t + \tanh^{-1} c).$$

1.14 Using the definitions of x and y in Solved Problem 1.2 and the definitions of A, B, and C in Solved Problem 1.3, compute

(a) Ax

(b) $x^T A$

(c) $x^T By$

(d) $x^T Cx$

(e) $Ax + 2Ay$

(f) $A(x + 2y)$

1.15 If A is an mxn matrix and x and y are n-dimensional vectors, show that $A(x+y) = Ax + Ay$.

1.16 Show that the criterion $J = \sum_{i=0}^{k-1} |x(i)| + \left(\sum_{i=k}^{N} |x(i)|^2 \right)^{\frac{1}{2}}$

possesses the Markovian property.

REFERENCES*

Introduction: [B-15], [B-32], [B-41], [B-65], [B-83], [L-14]

Vector-Matrix Notation: [B-91], [G-1]

Dynamical Systems: [B-32], [H-15], [K-7], [M-1], [Z-4]

Multistage Processes: [B-32], [B-65]

Multistage Decision Processes: [B-65], [H-3], [H-23], [K-13],
                                 [M-19], [N-4], [Y-1]

---

*The cited references represent a somewhat arbitrary selection chosen
to give the reader an entry into the literature. The complete bibliography should be consulted for many additional items.

Chapter 2

THE PRINCIPLE OF OPTIMALITY

AND DYNAMIC PROGRAMMING PROCESSES

INTRODUCTION

In this chapter we leave the confines of classical analysis
and enter the domain of dynamic programming. Our objective will be
to create an analytic structure within which we may study important
classes of multi-stage decision processes. As pointed out in the
previous chapter, in order to make any meaningful statements we
shall have to consider multi-stage decision processes possessing
certain desirable structural features. However, in return for
giving up extreme generality, we shall be able to shed significant
mathematical and computational light on numerous problems outside
the bounds of classical theory.

In pursuit of our goal of obtaining optimal policies, we shall
see that a basic principle, which Bellman has called "The Principle
of Optimality", is the key concept upon which our progress depends.
We shall also see how this principle can be related to some basic
equations which are then amenable to analytic and computational
study.

IMBEDDING AND RECURRENCE EQUATIONS

To introduce the dynamic programming ideas, consider the
problem of minimizing the separable cost function

$$J = \sum_{k=0}^{N} L(x(k), u(k), k) \qquad (2.1)$$

43

where x(0) has the fixed value c and where the system equation

$$x(k+1) = g(x(k),u(k),k), \quad k = 0,1,\ldots,N-1 \qquad (2.2)$$

and the constraints

$$x \in X \subset R^n \qquad\qquad\qquad (2.3)$$

$$u \in U \subset R^m \qquad\qquad\qquad (2.4)$$

must be satisfied. For simplicity, we shall assume that g and L are continuous functions of both arguments, that L is a positive bounded function of its arguments, and that both x and u range over closed, bounded subsets of R^n and R^m, respectively. The classical Weierstrass theorem then ensures that a minimizing policy exists.

The basic dynamic programming approach to the above problem is to regard it as a member of a class of similar problems and not as an isolated problem for initial stage 0 and initial state x(0) = c. In other words, we wish to imbed the problem we actually want to solve within a family of similar problems. The utility of this approach hinges upon our being able to find an imbedding for which:

(a) one member of the family of problems has a relatively
    simple solution, and
(b) relations may be obtained linking various members of
    the family of problems.
Providing steps (a) and (b) can be successfully negotiated, the solution of the original problem, which may be difficult to solve by itself, may be obtained by starting with the solution to the simple problem and then using the relations linking members of the family of problems to find the solution to the desired problem. Observe that this is the basic idea of mathematical induction as well as of numerous perturbation and continuity schemes of classical analysis. As with most such procedures, considerable skill is sometimes required to reformulate the statement of the original problem into an analytically tractable form.

To apply this approach to the above problem, observe that for fixed N the minimizing value of the function J depends upon only the two quantities k, the starting time, and x, the initial state. This observation suggests that we turn our attention to the family of problems associated with the minimization of

$$J(x,k) = \sum_{j=k}^{N} L(x(j), u(j), j) \qquad (2.5)$$

where now x(k) has the fixed value x, and where the system equations

$$x(j+1) = g(x(j), u(j), u), \quad j=k, k+1,\ldots,N-1, \qquad (2.6)$$

and the constraints

$$x \in X \subset R^n \qquad (2.7)$$
$$u \in U \subset R^m \qquad (2.8)$$

must be satisfied.

We first attempt to achieve step (b) above by finding a relation between members of this family of problems. In order to accomplish this goal, we define the minimum cost function I(x,k) as the minimum cost that can be obtained by using an admissible decision sequence for the remainder of the process starting from an arbitrary admissible state x∈X and an arbitrary stage k, $0 \leq k \leq N$. This function can be written

$$I(x,k) = \min_{u(k),u(k+1),\ldots,u(N)} \left\{ \sum_{j=k}^{N} L[x(j),u(j),j] \right\} \qquad (2.9)$$

where, since x is the state at stage k,

$$x(k) = x,$$

and where each decision u(j), u=k,...,N, must belong to the set U.

We then start the derivation of the desired equations by splitting the summation inside the braces into two parts: the term at stage k and the summation from stages k+1 through N. Formally,

$$I(x,k) = \min_{u(k),u(k+1),\ldots,u(N)} \left\{ L[x,u(k),k] \right.$$

$$\left. + \sum_{j=k+1}^{N} L[x(j),u(j),j] \right\} \qquad (2.10)$$

where the notation $x(k) = x$ is used in the first term inside the braces.

Next the minimization operation in Eq. (2.9) is also split· into two parts: a minimization over the current decision $u(k)$ and a minimization over the remaining decisions $u(k+1),u(k+2),\ldots u(N)$. In mathematical terms,

$$I(x,k) = \min_{u(k)} \min_{u(k+1),u(k+2),\ldots,u(N)} \left\{ L[x,u(k),k] \right.$$

$$\left. + \sum_{j=k+1}^{N} L[x(j),u(j),j] \right\} (2.11)$$

It can be seen that the first term in braces in Eq. (2.11) depends only on $u(k)$ and not on any $u(j)$, $j=k+1,\ldots,N$. Therefore, the minimization over $u(j)$, $j > k+1$, has no effect on this term, and

$$\min_{u(k)} \min_{u(k+1),u(k+2),\ldots,u(N)} \left\{ L[x,u(k),k] \right\} = \min_{u(k)} \left\{ L[x,u(k),k] \right\}$$

$$(2.12)$$

The second term in braces in Eq. (2.11) does not depend explicitly on $u(k)$. However, $u(k)$ does determine the state $x(k+1)$ through the state transformation equation

$$x(k+1) = g[x,u(k),k] . \qquad (2.13)$$

Using this relation and recalling the definition of the minimum cost function from Eq. (2.9), we have

$$\min_{u(k)} \quad \min_{u(k+1),u(k+2),\ldots,u(N)} \left\{ \sum_{j=k+1}^{N} L[x(j),u(j),j] \right\} =$$

$$\min_{u(k)} \left\{ I[g(x,u(k),k),\ k+1] \right\} \tag{2.14}$$

Combining Eqs. (2.12) and (2.14) into Eq. (2.11), and suppressing the index k on u(k), the functional equation for this problem can be written as

$$I(x,k) = \min_{u} \left\{ L[x,u,k] + I[g(x,u,k),\ k+1] \right\} . \tag{2.15}$$

This equation describes an iterative relation for determining $I(x,k)$ for all $x \epsilon X$ and for all k, $0 \le k \le N-1$, from knowledge of $I(x,k+1)$ for all $x \epsilon X$. This is the relation between members of the family of problems that we sought in (b) above.

In order to complete our approach, we need to satisfy item (a) above, i.e., we need to find one member of the family of problems that has a simple solution. If we evaluate the function $I(x,N)$ from Eq. (2.9), we see that

$$I(x,N) = \min_{u(N)} \quad \{L(x,u(N),N)\} \tag{2.16}$$

This determination for any $x \epsilon X$ is a minimization over the single variable $u(N)$, rather than the complex multistage minimization problem posed in (2.1). Thus, by first solving Eq. (2.16) for all $x \epsilon X$ and then applying the recursive equation, Eq. (2.15), we achieve the desired properties (a) and (b) noted above.

Note that the derivation of this equation has required virtually no assumptions regarding the nature of the functions in the problem formulation. In particular, the system difference equation $g(x,u,k)$ and the single-stage cost function $L(x,u,k)$ can vary with the stage and take on any nonlinear form. The

constraints sets X and U can also vary with the stage and be
nonlinear, and they can take on virtually any structure. We
could just as easily have used maximization for our basic operation
as minimization. (In this case, it would be more appropriate to
refer to the cost function as an objective function or performance
criterion. In general, we will use the term criterion function to
refer to either maximization or minimization.) Thus, we have been
able to obtain this extremely powerful result with little sacrifice
in generality.

We also see that a similar approach can be applied to any
criterion having the Markovian property. This further generalizes
the scope of this result. The next example illustrates this point.

EXAMPLE

2.1 Obtain recurrence formulas analogous to (2.15) for the
problems with cost functions

(a) $J = \prod_{i=0}^{N} L(x(i),u(i),i)$,

(b) $J = L(x(N),u(N),N)$,

using the system equations

$$x(k+1) = g(x(k),u(k),k)$$

and the constraints

$$x \in X \subset R^n, \quad u \in U \subset R^m$$

(a) Define

$$I(x,k) = \min_{u(k),u(k+1),\ldots,u(N)} \left\{ \prod_{i=k}^{N} L(x(i),u(i),i) \right\}$$

where $x(k) = x$. Arguing as above, we split the product into two
parts, a term for stage k and the other term present over stages

k+1,...,N. We further split the minimization into two parts, one
a minimization over u(k), the other a minimization over u(k+1),...,
u(N). The result is

$$I(x,k) = \min_{\substack{u \\ k=0,1,\ldots,N-1}} \{[L(x,u,k)]\ I[g(x,u,k),\ k+1]\}$$

$$I(x,N) = \min_{u} L(x,u,N)$$

(b) Define

$$I(x,k) = \min_{u(k),u(k+1),\ldots,u(N-1)} \{L(x,(N),u(N),N)\}$$

where x(k) = x. Arguing as above, we split the minimization into
two parts, one over u(k), the other over u(k+1),...,u(N). The
result is

$$I(x,k) = \min_{\substack{u \\ k=0,1,\ldots,N-1}} \{I(g(x,u,k),k+1)\}$$

$$I(x,N) = \min_{u} \{L(x,u,N)\}$$

The next few examples illustrate some variations of separable
criteria that can be handled in a straightforward manner.

EXAMPLE

2.2 Consider the problem with cost function

$$J = \psi[x(0)] + \sum_{k=1}^{N-1} [x^2(k) + u^2(k)] + \phi[x(N)],$$

system dynamics

$$x(k+1) = g(x(k),u(k),k), \quad k = 0,1,\ldots,N-1,$$

and constraints $x \in X \subset R^n$ and $u \in U \subset R^m$. This problem is the
same as that for which Eqs. (2.9) and (2.10) were derived, except
for two modifications. First, the initial state is not specified,
but instead it is to be chosen on the basis of an initial cost

function, $\psi[x(0)]$, that depends on the initial state $x(0)$. Second, the final decision does not affect the cost, but instead there is a terminal cost function, $\phi(x(N))$, that depends on the final state $x(N)$. Derive a set of recurrence relations for this problem.

At the final stage N, the terminal cost function is not affected by the final decision. Thus, the minimum cost function at this stage can be written directly in terms of the terminal cost function as

$$I(x,N) = \phi(x,N)$$

For intermediate stages, the recursive relation is as in Eq. (2.9), i.e.,

$$I(x,k) = \min_{u\varepsilon U} \{L(x,u,k) + I[g(x,u,k),\ k+1]\}$$

$$k = 1,2,\ldots,N-1$$

At the initial stage, the initial decision does not affect the initial cost function. However, it does affect the total cost by determining the state at stage 1 and the resulting cost of going to the end from there. The total cost for a given initial state x is thus

$$I(x,0) = \psi[x] + \min_{u\varepsilon U} \{I[g(x,u,0),1]\}$$

Finally, the optimum initial state is found by minimizing $I(x,0)$ over x. Formally,

$$\hat{x}(0) = \arg \min_{x} \{I(x,0)\}$$

where $\hat{x}(0)$ denotes the optimum initial state and arg denotes the operation of choosing the value of x that accomplishes the minimization of $I(x,0)$. (The notation arg is short for the word argument, since the operation performed is the determination of the argument of the function at its minimum.)

2.3   Consider the problem with cost function

$$J = \sum_{k=0}^{N-1} L(x(k),u(k),k),$$

system equation

$$x(k+1) = g(x(k),u(k),k),$$

and constraints $x \in X \subseteq R^n$, $u \in U \subseteq R^m$, $k=0,1,\ldots,K-1$.

(a)   Consider, in addition, the terminal constraint

$$x(N) = d,$$

where d is a single specified state.   Obtain a recurrence relation for this case.

(b)   Consider instead the terminal constraint

$$\psi(x(N)) = b$$

where ψ is a p-dimensional vector function, $1 \leq p \leq n-1$, and b is a p-dimensional vector of constants.   Obtain a recurrence relation for this case.

(a)   In this case there is only one admissible terminal state, d.   Therefore, at the final stage the minimum cost function is given by

$$I(x,N) = 0, \quad x = d$$

$$= \infty, \quad x \neq d$$

The recursive relation for other stages is as before, namely

$$I(x,k) = \min_{u \in U} \{L(x,u,k) + I[g(x,u,k),k+1]\}$$
$$k=0,1,\ldots,N-1$$

Clearly, the equation for $I(x,k)$ will always restrict our consideration to solutions where $x(N) = d$.   Computationally, this equation can be implemented either by setting $I(x,N)$ to an extremely large finite value for $x \neq d$, or by considering at each stage only states from which it is known that there exists a decision sequence that

takes us to state d at the final stage N. The relative advantages
of these methods will be made clear in later chapters on computational methods.

(b) In this case there is a family of admissible terminal states,
namely the set of terminal states that satisfy the equations
$\psi(x(N)) = b$. As in part (a), we calculate the minimum cost
function at the final stage as

$$I(x,N) = 0, \qquad \psi(x(N)) = b$$
$$= \infty, \qquad \psi(x(N)) \neq b$$

As above, the remaining iterative relations are

$$I(x,k) = \min_{u \varepsilon U} \{L(x,u,k) + I(g(x,u,k),k+1)\}$$
$$k=0,1,\ldots,N-1$$

The remarks about computational feasibility made in part (a)
apply here also.

2.4 Write the recurrence relation and obtain an explicit solution
for the problem with cost function

$$J = \sum_{k=0}^{2} [x^2(k) + u^2(k)] + x^2(3),$$

system equation

$$x(k+1) = x(k) + u(k)$$

and no constraints. (Since X and U are not bounded, we cannot
guarantee the existence of a solution directly from the Weierstrass theorem. Fortunately, however, other conditions are satisfied that guarantee the existence of a solution. These conditions
are discussed more fully in Volume 2 of this series.)

The recurrence relation can be written immediately as

$$I(x,k) = \min_{u} \{x^2 + u^2 + I(x + u, \ k+1)\}$$
$$k=0,1,2$$

$$I(x,3) = x^2$$

We begin by calculating $I(x,2)$

$$I(x,2) = \min_{u}\{x^2 + u^2 + (x+u)^2\}$$

Differentiating the quantity in brackets with respect to u and setting the result equal to 0, we have

$$\frac{d}{du} [x^2 + u^2 + (x+u)^2] = 2u + 2(x+u) = 2x + 4u = 0$$

We use this result and the convexity of the function $[x^2 + u^2 + (x+u)^2]$ to deduce that the minimizing value of u is $u = -\frac{1}{2} x$. Substituting into our equation for $I(x,2)$, we obtain

$$I(x,2) = x^2 + (-\frac{1}{2} x)^2 + (x - \frac{1}{2}x)^2 = \frac{3}{2} x^2$$

To obtain $I(x,1)$ we set up the recurrence relation as

$$I(x,1) = \min_{u} \{x^2 + u^2 + \frac{3}{2} (x+u)^2\}$$

By following the same procedure, we find that the minimizing value of u is $u = -\frac{3}{5} x$. Substituting this value into the recursive relation yields

$$I(x,1) = \frac{8}{5} x^2$$

Finally, the recursive relation for $I(x,0)$ is

$$I(x,0) = \min_{u} \{x^2 + u^2 + \frac{8}{5} (x+u)^2\}$$

We obtain the minimizing value of u as $u = -\frac{8}{13} x$ and the resulting minimum cost function as $I(x,0) = \frac{21}{13} x^2$.

The complete specification of $I(x,k)$, $k=0,1,2,3$, is thus as shown in Table 2.1 below.

Table 2.1   Explicit Solution for Minimum
Cost Function in Problem 2.4

k	$I(x,k)$
0	$1.615x^2$
1	$1.600x^2$
2	$1.500x^2$
3	x^2

2.5   Consider the case where the single-stage cost function $L(x,u,k)$
does not vary with the stage variable, i.e., $L(x,u,k) = L(x,u)$,
and where the system equation also does not vary with the stage
variable, i.e., $g(x,u,k) = g(x,u)$.

(a)   What form does the recurrence relation take?

(b)   In general, is the resulting minimum cost function $I(x,k)$ also
independent of the stage variable, i.e., does $I(x,k) = I(x)$?

(a)   We see directly that the equation for I becomes

$$I(x,k) = \min_{u} \{L(x,u) + I[g(x,u), k+1]\}$$
$$k=0,1,\ldots,N-1$$

$$I(x,N) = \min_{u}\{L(x,u)\}$$

(b)   Not always.   The preceding problem, Example 2.4, constitutes
a simple counter-example.   Conditions under which this result does
occur are discussed in Chapter 4.

BELLMAN'S PRINCIPLE OF OPTIMALITY

The recurrence relation discussed in the previous section is
the fundamental result of dynamic programming.   This relation is an
immediate consequence of a very important concept originated by

Bellman called the principle of optimality. This principle may
conceptually be thought of as follows: Given an optimal trajectory
from point A to point C, the portion of the trajectory from any
intermediate point B to point C must be the optimal trajectory from
B to C.

In Figure 2.1, if the path I-II is the optimal path from A to
C, then according to the principle of optimality path II is the
optimal path from B to C. The proof by contradiction for this case
is immediate: Assume that some other path, such as II', is the
optimum path from B to C. Then, path I-II' has less cost than path
I-II. However, this contradicts the fact that I-II is the optimal
path from A to C, and hence II must be the optimal path from B
to C.

Bellman states the principle of optimality formally, "An
optimal policy has the property that whatever the initial state and
initial decision are, the remaining decisions must constitute an
optimal policy with regard to the state resulting from the first
decision."

This principle is embodied in the recurrence relation Eq. (2.15);
the minimum cost from state x, stage k is found by minimizing the
sum of the current single-stage cost L(x,u,k) plus the minimum cost

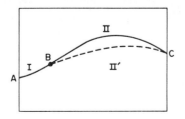

Figure 2.1   Illustration of the Principle of Optimality

of going to the end of the process from the resulting next state
g(x,u,k). Thus, we see that by carrying the minimum cost function
backward one stage at a time, we can find the minimum cost at any
state and stage by calculating one decision at a time.

It is clear, then, that the simple, intuitive concept inherent
in the principle of optimality provides the foundation for our
ability to derive recurrence relations of the type discussed in the
preceding section.

EXAMPLE

2.6 In Figure 2.1, is the portion of the optimal trajectory from
point A to point C that goes from point A to intermediate point B
the optimal trajectory from A to B?

_____

Yes. The proof by contradiction follows exactly as for the
path from B to C. In Chapter 4, the section Forward Dynamic
Programming will make clear the significance of this result.

THE OPTIMUM DECISION POLICY

Up to this point we have concerned ourselves only with deriving
recurrence relations for the minimum cost function, I(x,k). A very
natural question that arises now is how does one determine the
optimum sequence of decisions to achieve this minimum cost for a
given state and stage. One of the great strengths of dynamic
programming is that these sequences can be calculated with great
ease once the basic recurrence relation has been solved. To see
this, we need to develop the concept of an optimum decision policy.

Let us recall the form of Eq. (2.15), the basic recurrence
relation for I(x,k). A minimization over the admissible controls
uεU is required. Clearly, in the process of performing this minimi-

zation it should be straightforward to keep track of the values of
u that are being considered and to determine which value of u
actually leads to the minimum value. Let us denote the minimizing
values as $\hat{u}(x,k)$; formally,

$$\hat{u}(x,k) = \arg \min_{u} \{L[x,u,k] + I[g(x,u,k), k+1]\} \qquad (2.17)$$

where arg denotes the operation of specifying the minimizing value.
If we tabulate this value of $\hat{u}(x,k)$ for every state and stage, we
call the resulting function the <u>optimal decision policy</u>.

The optimal decision policy allows us to determine the optimum
decision sequence from any state and stage. Let us illustrate this
determination for a given state c at the initial stage 0. We first
utilize the optimal decision policy function to calculate the
optimum decision at this state and stage as $\hat{u}(c,0)$. Next, we
determine the next state by applying this decision, i.e., we calculate

$$\hat{x}(1) = g[c,\hat{u}(c,0),0] \qquad (2.18)$$

where $\hat{x}(1)$ is defined to be the state at stage 1 if we follow the
optimum sequence of decisions.

Now we utilize the optimum decision policy to evaluate
$\hat{u}(x(1),1)$; this decision now becomes the next decision in our
optimal sequence. The next step is to apply this decision from
state $\hat{x}(1)$ and calculate the next state, i.e., we determine

$$\hat{x}(2) = g[\hat{x}(1), \hat{u}(\hat{x}(1), 1), 1] \qquad (2.19)$$

where $\hat{x}(2)$ is the state at stage 2 if we follow the optimum
sequence of decisions.

The basic method for determining the optimum decision sequence
is now clear. Let us assume that by following the basic procedure
we have arrived at state $\hat{x}(k)$ at stage k. We then use the optimum
decision policy function to determine the next optimum decision in
our sequence $\hat{u}(\hat{x}(k),k)$. We then determine the next state as

$$\hat{x}(k+1) = g[\hat{x}(k),\hat{u}(\hat{x}(k), k), k] \qquad\qquad (2.20)$$

This procedure begins at the initial state c at stage 0 and
continues until we read the final state at the final stage N.
Note that we obtain both the optimum sequence of decisions,
$\hat{u}(\hat{x}(k),k)$, k=0,1,...,N, and the resulting optimum sequence of
states, $\hat{x}(k)$, k=0,1,...,N, where $\hat{x}(0) = c$. We call the latter
sequence of states the optimal trajectory in state space or, more
simply, the optimal trajectory.

It is clear that the above procedure can equally well be applied
at any state and stage, not just at a fixed initial state and stage.
All that is required is that we begin with the given state and
stage. The fact that we can find an optimum decision sequence from
any initial state and stage further increases the power and scope
of dynamic programming; in the course of solving one problem we
have truly solved a very large number of similar problems.

It is also clear that the procedure can be applied when the
initial state is not specified but must be determined by minimizing
a function over all possible initial states (see Problem 2.2). In
this case, the initial state is first selected by performing the
minimization, and the optimum decision sequence and optimal
trajectory from this state are recovered by the usual procedure.

EXAMPLE

2.7 Assume that the optimum decision policy is given by $\hat{u}(x,k) =$
$\phi(k)x$, k=0,1,...,9. Assume also that the system equation is given by

$$x(k+1) = x(k) + u(k)$$

What is the optimum decision sequence and optimal trajectory from state x=1 and stage k=6?

The solution is summarized in the tables below.

Table 2.2   Optimal Trajectory from x=1, k=6

k	OPTIMAL TRAJECTORY, $\hat{x}(k)$
6	1
7	$1 + \phi(6)$
8	$1 + \phi(6) + \phi(7) + \phi(7)\phi(6)$
9	$1+\phi(6)+\phi(7)+\phi(8)+\phi(7)\phi(6)+\phi(8)\phi(6)+\phi(8)\phi(7)+\phi(8)\phi(7)\phi(6)$

Table 2.3   Optimal Decision Sequence from x=1, k=6

	OPTIMAL DECISION SEQUENCE, $\hat{u}(\hat{x}(k),k)$
6	$\phi(6)$
7	$\phi(7) [1 + \phi(6)]$
8	$\phi(8) [1 + \phi(6) + \phi(7) + \phi(7)\phi(6)]$
9	$\phi(9) [1 + \phi(6) + \phi(7) + \phi(8) + \phi(7)\phi(6) + \phi(8)\phi(6) + \phi(8)\phi(7) + \phi(8)\phi(7)\phi(6)]$

EXAMPLE

2.8   Refer to Problem 2.4, where the minimum cost function is found explicitly.

(a)   Calculate the corresponding optimum decision policy function.

(b)   Calculate the optimum sequence of decisions and the optimal trajectory from x(0) = 1.

(a)   Since no decision is made at stage 3, $\hat{u}(x,3)$ has no meaning.
In the calculations for $I(x,2)$, we see that the minimizing u for any
x is $-\frac{1}{2}$ x.   Therefore,

$$\hat{u}(x,2) = -\frac{1}{2} x.$$

In the calculation for $I(x,1)$ we see that the minimizing u for any
x is $-\frac{3}{5}$ x.   Therefore

$$\hat{u}(x,1) = -\frac{3}{5} x.$$

In the calculation for $I(x,0)$ we see that the minimizing u for any
x is u $= -\frac{8}{13}$ x.   Therefore,

$$\hat{u}(x,0) = -\frac{8}{13} x$$

These results are summarized in Table 2.4 below.

Table 2.4   Optimum Decision Policy Function for Problem 2.4

k	$\hat{u}(x,k)$
0	−0.615x
1	−0.600x
2	−0.500x
3	

(b)   The optimum decision sequence and optimal trajectory from
$x(0) = 1$ can be calculated directly as in Table 2.5 below.

Table 2.5   Optimum Decision Sequence and Optimal Trajectory
from $x(0) = 1$ in Problem 2.4

k	Optimal Trajectory, $\hat{x}(k)$	Optimal Decision Sequence $\hat{u}(x(k),k)$
0	1.000	−0.615
1	0.385	−0.231
2	0.154	−0.077
3	0.077	

We have thus seen that the dynamic programming method involves two sweeps through the stage variable. In the first sweep we are working backwards, computing $I(x,k)$, the minimum cost function at stage x and stage k, in terms of $I(x,k+1)$, the minimum cost function at stage k+1. In the second sweep we are working forward, recovering the optimum decision sequence and the optimal trajectory in state space by forward iteration of the system equation $\hat{x}(k+1) = g[\hat{x}(k), \hat{u}\{\hat{x}(k),k\},k]$, where the optimum decision policy function $\hat{u}(\hat{x}(k),k)$ is used to determine the optimum decision at each stage. The computational implications of these results will be made clear in Chapter 3 of this volume.

SUMMARY

We have now developed the basic theoretical tools of dynamic programming and are in a position to apply it to a variety of problems. As we have seen, the theory rests in a single concept of great power and simplicity -- Bellman's Principle of Optimality. This principle can be stated in a number of ways; basically, it tells us that an optimum decision policy has the property that any portion of an optimal trajectory from an intermediate state to the final state is itself the optimal trajectory from the intermediate state. This allows us to determine a total optimum decision policy and the corresponding minimum cost function by starting at the end of the process and working backward one stage at a time, considering only the decision at that stage. In determining this decision, we must consider both the short-term cost at that stage and the long-term consequences of having to follow the optimal policy from the next state to which this decision takes us. After having made this sweep backward through the stages, we can determine the optimum decision sequence and optimal trajectory in state space for any initial state by sweeping forward with the system equations, using the optimum decision policy at each stage to determine the next decision. As we shall see in the remainder of this book and in the next volume, this approach to solving problems can be applied to obtaining analytic and computational results of great importance in numerous areas.

SOLVED PROBLEMS

2.1 Use dynamic programming to derive recursive equations for the
minimax criterion. In this case, the objective is to minimize the
maximum value that a particular function takes on during the
decision interval. This type of problem is relevant to worst-case
analyses of complex systems. Mathematically, we seek to minimize
the performance criterion

$$J = \max_{k=0,1,\ldots,N} [L(x(k),u(k),k)]$$

where the system equation is

$$x(k+1) = g(x(k),u(k),k)$$

For this case we proceed in the usual manner by defining

$$I(x,k) = \min_{u(k),\ldots,u(N)} \left[\max_{j=k,\ldots,N} \{L(x(j),u(j),j)\} \right]$$

where $x(k) = x$. We obtain immediately

$$I(x,k) = \min_{u} \left[\max \left\{ L(x,u,k), \ I[g(x,u,k), \ k+1] \right\} \right]$$

The starting condition is

$$I(x,N) = \min_{u} \{L(x,u,N)\}$$

2.2 Use dynamic programming to derive recursive equations for the
"product of sums" criterion. This criterion is appropriate to certain types of resource allocation problems (see Problem 2.5). Consider here the case where the criterion to be maximized takes the form

$$J = \prod_{i=1}^{N} \left(\sum_{j=1}^{K_i} L_{ij}(u_{ij}) \right)$$

subject to the constraints

$$0 \leq u_{ij} \leq \Omega_{ij}, \; j=1,\ldots,K_i, \; i=1,\ldots,N$$

$$0 \leq \sum_{j=1}^{K_i} u_{ij} \leq \Omega_i, \; i=1,\ldots,N$$

$$0 \leq \sum_{i=1}^{N} \sum_{j=1}^{K_i} u_{ij} \leq \Omega$$

All of the functions L_{ij} are assumed to be nonnegative and mono-tonically nondecreasing functions of their respective arguments u_{ij}, with Ω_{ij}, Ω_i, and Ω being constants.

For this case we first use the nonnegativity and monotonicity of the L_{ij} to re-write J as

$$J = \prod_{i=1}^{N} \Lambda_i \, (y_i)$$

where

$$\Lambda_i(y_i) = \max_{\substack{0 \leq u_{ij} \leq \Omega_{ij} \\ j=1,\ldots,K_i}} \left\{ \sum_{j=1}^{K_i} L_{ij}(u_{ij}) \right\}$$

and where

$$y_i = \sum_{j=1}^{K_i} u_{ij}.$$

The nonnegativity of the L_{ij} implies that each Λ_i is nonnegative; hence, for a given y_i the product J will be maximized by choosing the maximum possible $\Lambda_i(y_i)$. The monotonicity of the L_{ij} implies that in maximizing Λ_i the equality condition on the y_i will be met.

We note that the $\Lambda_i(y_i)$ can be determined by solving recurrence relations. We define

$$I_i^*(x,k) = \max_{\substack{0 \le u_{ij} \le \Omega_{ij} \\ j=k,\ldots,K_i}} \left\{ \sum_{j=k}^{K_i} L_{ij}(u_{ij}) \right\}$$

where

$$x = \sum_{j=k}^{K_i} u_{ij}$$

The latter equation can be converted to system equation form by writing

$$x(k) = \sum_{j=k}^{K_i} u_{ij}$$

It follows immediately that

$$x(k+1) = x(k) - u_{ik}$$

The desired equation thus becomes

$$I_i^*(x,k) = \max_{0 \le u_{ik} \le \Omega_{ik}} \left\{ L_{ik}(u_{ik}) + I_i^*(x-u_{ik}, k+1) \right\}$$

with system equation

$$x(k+1) = x(k) - u_{ik}.$$

The starting condition is

$$I_i^*(x,K) = L_{iK}(u_{iK}).$$

The desired function $\Lambda_i(y_i)$ is obtained as

$$\Lambda_i(y_i) = I_i^*(y_i, 1),$$

where the range of y_i is

$$0 \le y_i \le \Omega_i$$

We then define

$$I(x,k) = \max_{\substack{0 \le y_i \le \Omega_i \\ i=k,\ldots N}} \left\{ \prod_{i=k}^{N} \Lambda_i(y_i) \right\}$$

subject to

$$x = \sum_{i=k}^{N} y_i$$

This equality constraint can also be converted to our usual system equation form by defining

$$x(k) = \sum_{i=k}^{N} y_i$$

It follows immediately that

$$x(k+1) = x(k) - y_k.$$

We thus obtain the desired recurrence relation as

$$I(x,k) = \max_{0 \le y_k \le \Omega_k} \{\Lambda_k(y_k) \, I(x-y_k, \, k+1)\}$$

subject to

$$x(k+1) = x(k) - y_k$$

The starting condition is

$$I(x,N) = \max_{0 \le y_N \le \Omega_N} \{\Lambda_N(y_N)\}$$

The desired value of J is

$$J = I(\Omega,1)$$

The complete solution is thus obtained by solving N recursive equations of the summation criterion type to obtain $\Lambda_i(y_i)$ = $I_i^*(y_i,1)$ and then solving one recursive equation of the product type to obtain $J = I(\Omega,1)$.

2.3 Consider the problem with system equation

$$x(k+1) = x(k) + u(k),$$

performance criterion

$$J = \sum_{k=0}^{2} [x^2(k) + u^2(k)],$$

and constraints

$$0 \leq x \leq 4$$
$$-1 \leq u \leq 1$$

Find the minimum cost function and optimum decision policy as analytical functions of x for k = 0,1,2.

We note that at stage 2,

$$\hat{u}(x,2) = 0$$
$$I(x,2) = x^2$$

We then examine the recursive equation for stage 1

$$I(x,1) = \min_{u} \{x^2 + u^2 + I(x + u, 2)\}$$

$$= \min_{u} \{x^2 + u^2 + (x+u)^2\} \quad .$$

If we assume that the constraints do not apply and differentiate as in Problem 2.4 in the text, we obtain

$$\hat{u}(x,1) = -\frac{1}{2} x$$

$$I(x,1) = \frac{3}{2} x^2 \quad .$$

Clearly, for $0 \leq x \leq 2$, it follows that $-1 \leq \hat{u} \leq 0$, and the constraint does not apply. Therefore, the above solution does indeed hold for $0 \leq x \leq 2$.

Now, for $x > 2$, considering the constraint $-1 \leq u \leq 0$, we see that the quantity in brackets is minimized if $\hat{u} = -1$. Therefore, for $x > 2$,

$$\hat{u}(x,1) = -1$$
$$I(x,1) = x^2 + (-1)^2 + (x-1)^2 = 2x^2 - 2x + 2$$

To summarize,

$$\hat{u}(x,1) = \begin{cases} -\dfrac{1}{2} x \quad , & 2 \geq x \geq 0 \\[2ex] -1 & 4 \geq x \geq 2 \end{cases}$$

$$I(x,1) = \begin{cases} \dfrac{3}{2} x^2 \, , & 2 \geq x \geq 0 \\[2ex] 2x^2 - 2x + 2 & 4 \geq x \geq 2 \end{cases}$$

Now let us examine stage 0. The recursive equation is

$$I(x,0) = \min_{u} \{x^2 + u + I(x+u, 1)\} \ .$$

Let us first assume that $0 \leq x + u \leq 2$. Then

$$I(x,0) = \min_{u} \{x^2 + u^2 + \frac{3}{2} (x+u)^2\} \ .$$

The solution is

$$\hat{u}(x,0) = -\frac{3}{5} x$$

$$I(x,0) = 1\frac{3}{5} x^2$$

Now, $0 \leq x + \hat{u}(x,0) \leq 2$ only if $0 \leq \frac{2}{5} x \leq 2$, or if $0 \leq x \leq 5$. Since this covers the region $0 \leq x \leq 3$, we see that if $-1 \leq \hat{u}(x,0) \leq 0$ for $0 \leq x \leq 3$, then the above expression is exact. We see, however, that it holds only for $0 \leq x \leq \frac{5}{3}$.

On the other hand, we see that for $x \geq \frac{5}{3}$, $\hat{u}(x,0) = -1$. We first assume $I(x-1,1) = \frac{3}{2}(x-1)^2$. In that case

$$I(x,0) = x^2 + (-1)^2 + \frac{3}{2}(x-1)^2 = \frac{5}{2}x^2 - 3x + \frac{5}{2}.$$

We see that for $\frac{5}{3} \leq x \leq 3$, $\frac{2}{3} \leq x-1 \leq 2$. Therefore, for this range of x, the next state falls in the region where $I(x,1) = \frac{3}{2}x^2$, and the above expression is correct.

For $x \geq 3$, we see that $x-1 \geq 2$, so that $I(x-1,1)$ is given by $[2(x-1)^2 - 2(x-1) + 2]$. Therefore, for $3 \leq x \leq 4$,

$$\hat{u}(x,0) = -1$$
$$I(x,0) = x^2 + (-1)^2 + 2(x-1)^2 - 2(x-1) + 2 = 3x^2 - 6x + 7$$

Thus, we see that

$$u(x,0) = \begin{cases} -\dfrac{3}{5}x & , \quad 0 \leq x \leq \dfrac{5}{3} \\[3mm] -1 & , \quad \dfrac{5}{3} \leq x \leq 4 \end{cases}$$

$$I(x,0) = \begin{cases} \dfrac{8}{5}x^2 & , \quad 0 \leq x \leq \dfrac{5}{3} \\[3mm] \dfrac{5}{2}x^2 - 3x + \dfrac{5}{2} , \quad \dfrac{5}{3} \leq x \leq 3 \\[3mm] 3x^2 - 6x + 7 , \quad 3 \leq x \leq 4 \end{cases}$$

2.4 Consider the problem with the following performance criterion which is to be minimized

$$J = e^{3u^2(0)} e^{2u^2(1)} e^{u^2(2)} \cosh[u(3)] \, e^{3x^2(0)} e^{2x^2(1)} e^{x^2(2)} e^{x^2(3)}.$$

The system equation is

$$x(k+1) = x(k) + u(k) ,$$
$$k = 0, 1, 2 \quad ,$$

and there are no constraints.

(a) Find the <u>complete</u> dynamic programming solution to this problem, including both the minimum cost function and the optimum decision function.

(b) Recover the optimal trajectory in state space and the optimum decision sequence from $x(0) = 1$.

NOTE: $\cosh u = \frac{1}{2}(e^u + e^{-u})$

(a) To solve the problem, we first start at the last control, $u(3)$. Since $x(4)$ does not appear in the criterion, we can minimize $\cosh [u(3)]$ directly. We see that the minimizing u satisfies

$$\frac{d}{du} [\cosh u] = \frac{1}{2} e^u - \frac{1}{2} e^{-u} = 0 \, ,$$

so that $\cosh u$ is minimized at $u = 0$. Therefore,

$$\hat{u}(x,3) = 0$$

We then see that since $\cosh (0)=1$, we are left with the minimization of

$$J^* = e^{3u^2(0)} e^{2u^2(1)} e^{u^2(2)} e^{3x^2(0)} e^{2x^2(1)} e^{x^2(2)} e^{x^2(3)}$$

We can either work <u>directly</u> with this formula or else note that the values of $u(0)$, $u(1)$ and $u(2)$ that minimize J^* will minimize $\ln J^*$. Let us use the direct approach. We see first that

$$I(x,3) = e^{x^2}$$

Next, we try to find $u(x,2)$ and $I(x,2)$. We write

$$I(x,2) = \min_u \left\{ e^{x^2} e^{u^2} e^{(x+u)^2} \right\}$$

If we differentiate with respect to u, we find the minimizing u solves

$$\frac{d}{du}\left(e^{x^2}e^{u^2}e^{(x+u)^2}\right) = e^{x^2}e^{u^2}e^{(x+u)^2}[2u + 2(x+u)] = 0.$$

Clearly, this derivative is zero when $u = -\frac{1}{2}x$. Therefore,

$$\hat{u}(x,2) = -\frac{1}{2}x$$

By substitution,

$$I(x,2) = e^{x^2}e^{(-\frac{1}{2}x)^2}e^{(x-\frac{1}{2}x)^2} = e^{\frac{3}{2}x^2} = I(x,2)$$

Next, we find $\hat{u}(x,1)$ and $I(x,1)$ from

$$I(x,1) = \min_u \left\{ e^{2x^2}e^{2u^2}e^{\frac{3}{2}(x+u)^2} \right\}$$

Differentiating, we obtain

$$\frac{d}{du}\left(e^{2x^2}e^{2u^2}e^{\frac{3}{2}(x+u)^2}\right) = e^{2x^2}e^{2u^2}e^{\frac{3}{2}(x+u)^2}[4u + 3(x+u)] = 0.$$

We obtain

$$\hat{u}(x,1) = -\frac{3}{7}x$$

$$I(x,1) = e^{2x^2}e^{\frac{18}{49}x^2}e^{\frac{3}{2}\cdot\frac{16}{49}x^2} = e^{\frac{20}{7}x^2} = I(x,1)$$

Finally, we find $\hat{u}(x,0)$ and $I(x,0)$ from

$$I(x,0) = \min_u \left\{ e^{3x^2}e^{3u^2}e^{\frac{20}{7}(x+u)^2} \right\}$$

Differentiating, we obtain

$$\frac{d}{dx}\left(e^{3x^2}e^{3u^2}e^{\frac{20}{7}(x+u)^2}\right) = e^{3x^2}e^{3u^2}e^{\frac{20}{7}(x+u)^2}\left[6u + \frac{40}{7}(x+u)\right] = 0.$$

We see that

$$\hat{u}(x,0) = -\frac{20}{41} x$$

$$I(x,0) = e^{3x^2} e^{\frac{3\cdot20\cdot20}{41\cdot41} x^2} e^{\frac{20}{7}\cdot\frac{21}{41}\cdot\frac{21}{41} x^2} = e^{\frac{183}{41} x^2} = I(x,0)$$

(b)   To obtain the optimal trajectory in state space and the optimum decision sequence from $x(0) = 1$, we utilize $\hat{u}(x,k), k = 0,1,2,3$, generating

k	$\hat{x}(k)$	$\hat{u}(k)$
0	1	$-\dfrac{20}{41}$
1	$\dfrac{21}{41}$	$-\dfrac{9}{41}$
2	$\dfrac{12}{41}$	$-\dfrac{6}{41}$
3	$\dfrac{6}{41}$	0

2.5   Consider the following general resource allocation problem. Assume that a fixed resource can be used for any of N tasks. Let the return from using y units of resource on task k, k=1,2,...,N be defined by

Return = $R(y,k)$

where the amount of resource utilized on task k is bounded by $0 \leq y \leq Y(k)$. Develop a dynamic programming equation to find the best utilization of B units of resource.

The dynamic programming formulation can be established as follows. Let

x(k) = amount of resource left to utilize on tasks k, k+1,...K.

u(k) = amount of resource used on task k.

Then the system equation becomes

$$x(k+1) = x(k) - u(k)$$

i.e., the resource left to utilize on tasks k+1,...,K is the amount left to use on k,...,N less the amount used on k. The performance criterion, which is to be maximized, is the total return, given by

$$J = \sum_{k=1}^{N} R(u,k).$$

The constraints are:

$$0 \leq u(k) \leq Y(k)$$

$$0 \leq x(k) \leq B$$

The iterative equation becomes

$$I(x,k) = \max_{0 \leq u \leq Y(k)} \left\{ R(u,k) + I(x-u, k+1) \right\}$$

where the boundary condition is

$$I(x,N) = R(x,N)$$

i.e., if there is any resource remaining at the last task, it will all be used on this task. The best utilization of B units of resource over the N tasks is found by applying the above recursive equation to find I(B,1) and then tracing out the solution from this state and stage.

2.6 A simplified power system consists of three generating units that meet a given demand, as shown in Figure 2.2. The output of unit i is x_i, in megawatts (MW), i=1,2,3. The operating cost over one time increment of unit i is C_i, in dollars, i=1,2,3, where $C_1 = \frac{1}{2} x_1^2$, $C_2 = x_2^2$, and $C_3 = \frac{3}{2} x_3^2$. Unit 1 is a must-run unit and is on at all times. Units 2 and 3 can either be on or off; however, unit 2 will always be started up before unit 3. Let us now apply dynamic programming to solve the <u>economic dispatch</u> problem of electric utility operation. If one unit is operating and the demand is D, then, provided that the capacity of the unit exceeds the demands, it is clear that the system operating cost is $f_1(D) = \frac{1}{2}D^2$.

(a) If two units are on, and the demand is D, apply dynamic programming to find the <u>minimum</u> system operating cost as $f_2(D)$. Assume that D is such that neither unit is capacity limited.

(b) If all three units are on, and the demand is D, apply dynamic programming to find the <u>minimum</u> system operating cost as $f_3(D)$. Assume that D is such that no unit is capacity limited.

The economic dispatch problem can be viewed as a resource allocation problem, where the demand is distributed over the units to minimize cost. The quantities $f_i(D)$, $1 = 1,2,3$, are minimum cost functions for meeting a given demand D with i units on. Note that $f_i(D) \leq f_j(D)$ if $i < j$, since the flexibility of having an additional unit to generate power can only decrease the total cost of meeting a specific demand D.

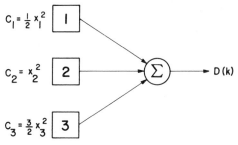

Figure 2.2   Simplified Power Systems

(a) We note from the priority order that if two units are on, they must be units 1 and 2. Following our resource allocation termino-logy, let us define

$x(k)$ = amount of demand furnished by units k,...,2.

(k = 1 or 2)

$u(k)$ = amount of demand furnished by unit k

$R(u(k))$ = cost of furnishing demand u with unit k.

We then have a two-stage dynamic programming problem, where the system equations are

$$x(2) = x(1) - u(1)$$
$$x(1) = D,$$

and where

$$R(u(k)) = \tfrac{1}{2}u^2(1), \quad k=1$$
$$= u^2(2), \quad k=2.$$

We see immediately that

$$I(x,2) = x^2,$$

since all demand x must be furnished by unit 2. By assumption, x is less than the capacity of unit 1. Next we can show that

$$I(x,k) = \min_{u}\{\tfrac{1}{2}u^2 + (x-u)^2\}.$$

Differentiating with respect to u,

$$\frac{\partial I}{\partial u} = 0 = u - 2(x-u)$$

from which we see that

$$\hat{u} = \frac{2}{3}x,$$

as long as demand is such that neither unit is forced to work above its capacity. Then we see that

$$I(x,1) = \frac{1}{2}\left(\frac{2}{3}x\right)^2 + \left(\frac{1}{3}x\right)^2$$

$$= \frac{1}{3}x^2.$$

By interpreting $I(x,1)$ as the minimum cost of meeting demand x with the two units on, we obtain

$$f_2(D) = \frac{1}{3}D^2.$$

(b) We could use a procedure analogous to that in part (a) and
define a three-stage resource allocation problem. However, it is
more efficient to note that $f_2(D)$ is the minimum cost of meeting a
given demand D with units 1 and 2 on. Thus, we can write immediately

$$f_3(D) = \min_u \{ \frac{3}{2} u^2 + f_2 (D-u)^2 \}$$

$$= \min_u \{ \frac{3}{2} u^2 + \frac{1}{3} (D-u)^2 \}$$

Differentiating,

$$\frac{\partial f_3}{\partial u} = 0 = 3u - \frac{2}{3} (D-u),$$

implying

$$\hat{u} = \frac{2}{11} D$$

provided no unit is forced to operate above its capacity. Substituting,

$$f_3(D) = \frac{3}{2} (\frac{2}{11} D)^2 + \frac{1}{3} (\frac{9}{11} D)^2$$

$$= (\frac{6}{121} + \frac{27}{121}) D^2$$

$$= \frac{3}{11} D^2 .$$

It is interesting to note that, in general,

$$f_n(D) = \min_u \{C_n (u) + f_{n-1} (D-u)\}$$

where $C_n(u)$ is the cost of meeting demand u with unit n. The minimum cost functions, $f_n(D)$, are called composite cost functions for
n units being on, and they are widely-used functions in electric
utility operation. Dynamic programming thus provides an efficient
recursive relation for computing these functions.

2.7 A fundamental problem of aeronautical and aerospace engineering
is the determination of optimal trajectories for powered flight
vehicles. The vehicle may be an aircraft, missile, or satellite.
The cost function may be minimum fuel, minimum time, minimum energy,
or some other function related to the performance of the vehicle.
Numerous constraints corresponding to the physical limitations of
the vehicle must be obeyed.

A typical problem of this type is shown in Figure 2.3. The
vehicle, of constant mass M, is assumed to be flying in two spatial
dimensions, altitude z and ground track y. It is assumed that all
external forces on the vehicle, including all aerodynamic forces
(lift, drag) and all gravitational forces, can be resolved into
components E_y and E_z, along the y-axis and z-axis, respectively.
Furthermore, it is assumed that the propulsion of the powered
vehicle (jet engine, rocket engine, reaction jets, etc.) and the
control surfaces (elevators, ailerons, etc.) are such that the
control forces on the vehicle can be resolved into components C_y
and C_z, along the y-axis and z-axis respectively. These control
forces are assumed to be constrained by the relations

$$\alpha_1^- \leq C_z \leq \alpha_1^+$$

$$\alpha_2^- \leq C_y \leq \alpha_2^+$$

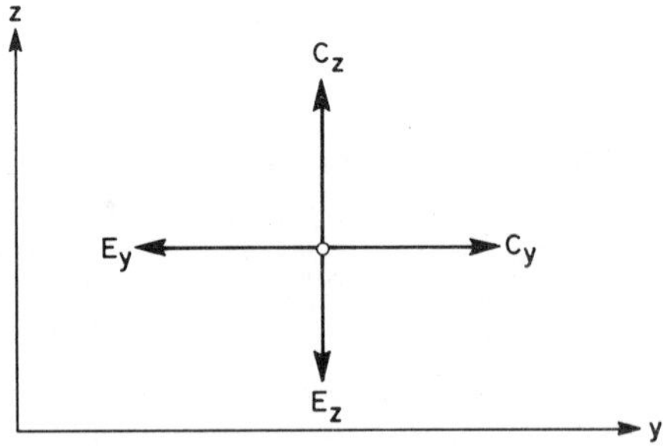

Figure 2.3   Forces on Vehicle in Powered Flight Control
              Problem

The altitude of the aircraft is constrained by

$$\beta_1^- \leq z \leq \beta_1^+$$

The ground track is restricted to lie in the interval

$$\beta_2^- \leq y \leq \beta_2^+$$

The vertical and horizontal components of velocity are constrained by

$$\beta_3^- \leq \dot{z} \leq \beta_3^+$$

$$\beta_4^- \leq \dot{y} \leq \beta_4^+$$

The vertical and horizontal components of acceleration are constrained by

$$\beta_5^- \leq \ddot{z} \leq \beta_2^+$$

$$\beta_6^- \leq \ddot{y} \leq \beta_2^+$$

The external forces on the vehicle are known functions of the position and velocity coordinates and the control forces

$$E_z = f_1(z,y,\dot{z},\dot{y},C_z, C_y)$$

$$E_y = f_2(z,y,\dot{z},\dot{y},C_z, C_y)$$

The cost criterion, which is to be minimized, has the form

$$J = \int_{t_o}^{t_f} L(z,y,\dot{z},\dot{y},C_z,C_y,t) \ dt$$

where t_o is the specified initial time and t_f is a specified terminal time.

At the initial time t_o the vehicle is known to be at altitude z_o and ground track y_o with vertical velocity $\dot{z}_o$ and horizontal velocity $\dot{y}_o$. At the terminal time t_f the vehicle is constrained to

be at a specified altitude z_f and ground track y_f with specified
vertical velocity $\dot{z}_f$ and horizontal velocity $\dot{y}_f$.

(a) Formulate this problem in terms of state-variable notation

(b) Apply the Euler integration formula,

$$\int_{t^*}^{t^* + \Delta t} h(t) \ dt \cong h(t^*) \ \Delta t,$$

to obtain a discrete time-version of the problem. Use uniform
quantization increments in time of

$$\Delta t = \frac{t_f - t_o}{N} \ .$$

(c) Derive a dynamic programming recurrence relation for this
discrete-time problem.

(a) We first note that it will be necessary to carry the two
position variables z and y and the two corresponding velocity
variables $\dot{z}$ and $\dot{y}$ as state variables. The forces C_z and C_y are
clearly control variables. The other two candidates for state
variables are accelerations $\ddot{z}$ and $\ddot{y}$. However, by applying Newton's
second law on the net force in the z direction, we see that

$$F_z = M\ddot{z} = C_z - E_z$$

from which we deduce that

$$\ddot{z} = \frac{1}{M} [C_z - f_1(z,y,\dot{z},\dot{y},C_z,C_y)]$$

A similar analysis in the y direction shown that

$$\ddot{y} = \frac{1}{M} [C_y - f_2(z,y,\dot{z},\dot{y},C_z,C_y)]$$

These two equations allow us to replace the acceleration variables
by known functions of the position and velocity variables.

We now define the four-dimensional state vector x as

$$
x = \begin{bmatrix} x_1 \\ x_2 \\ x_3 \\ x_4 \end{bmatrix} = \begin{bmatrix} z \\ y \\ \dot{z} \\ \dot{y} \end{bmatrix}
$$

The two-dimensional control vector u becomes

$$
u = \begin{bmatrix} u_1 \\ u_2 \end{bmatrix} = \begin{bmatrix} C_z \\ C_y \end{bmatrix}
$$

The continuous-time state equation becomes

$$
\dot{x} = \begin{bmatrix} \dot{x}_1 \\ \dot{x}_2 \\ \dot{x}_3 \\ \dot{x}_4 \end{bmatrix} = \begin{bmatrix} x_3 \\ x_4 \\ \frac{1}{M} \left[u_1 - f_1(x_1,x_2,x_3,x_4,u_1,u_2) \right] \\ \frac{1}{M} \left[u_2 - f_2(x_1,x_2,x_3,x_4,u_1,u_2) \right] \end{bmatrix} = f(x,u)
$$

The cost function becomes

$$
J = \int_{t_o}^{t_f} L(x_1,x_2,x_3,x_4,u_1,u_2)dt = \int_{t_o}^{t_f} L(x,u)dt
$$

The constraints take the form

$$
\alpha_1^- \le u_1 \le \alpha_1^+
$$

$$
\alpha_2^- \le u_2 \le \alpha_2^+
$$

$$
\beta_1^- \le x_1 \le \beta_1^+
$$

$$\beta_2^- \le x_2 \le \beta_2^+$$

$$\beta_3^- \le x_3 \le \beta_3^+$$

$$\beta_4^- \le x_4 \le \beta_4^+$$

$$\beta_5^- \le u_1 - f_1(x_1,x_2,x_3,x_4,u_1,u_2) \le \beta_5^+$$

$$\beta_6^- \le u_2 - f_2(x_1,x_2,x_3,x_4,u_1,u_2) \le \beta_6^+$$

The terminal constraint is

$$x(t_f) = \begin{bmatrix} x_1(t_f) \\ x_2(t_f) \\ x_3(t_f) \\ x_4(t_f) \end{bmatrix} = \begin{bmatrix} z_f \\ y_f \\ \dot{z}_f \\ \dot{y}_f \end{bmatrix} = x_f$$

The initial condition is

$$x(t_o) = \begin{bmatrix} x_1(t_o) \\ x_2(t_o) \\ x_3(t_o) \\ x_4(t_o) \end{bmatrix} = \begin{bmatrix} z_o \\ y_o \\ \dot{z}_o \\ \dot{y}_o \end{bmatrix} = x_o$$

(b)   We make the correspondence between the index k, k=0,1,...,N, and the N+1 quantized time instants between t_o to t_f.

$$k \stackrel{\Delta}{=} t_o + k\Delta t$$

We define

$$x(k) \stackrel{\Delta}{=} x(t_o + k\Delta t)$$

$$u(k) \stackrel{\Delta}{=} u(t_o + k\Delta t).$$

We then apply the Euler integration formula to rewrite the cost function as

$$J = \sum_{k=0}^{N-1} L(x(k),\ u(k))\Delta t$$

$$= \sum_{k=0}^{N-1} L(x_1(k),\ x_2(k),\ x_3(k),\ x_4(k),\ u_1(k),\ u_2(k))\Delta t$$

The Euler integration formula can also be rewritten as

$$x(k+1) = x(k) + f(x(k),\ u(k))\Delta t$$

or, in expanded form,

$$x(k+1) = \begin{bmatrix} x_1(k+1) \\ x_2(k+1) \\ x_3(k+1) \\ x_4(k+1) \end{bmatrix} =$$

$$
= \begin{bmatrix}
x_1(k) + x_3(k)\Delta t \\[2ex]
x_2(k) + x_4(k)\Delta t \\[2ex]
x_3(k) + \dfrac{1}{M}\left(u_1(k) - f_1[x_1(k),x_2(k),x_3(k),x_4(k),u_1(k),u_2(k)]\right)\Delta t \\[2ex]
x_4(k) + \dfrac{1}{M}\left(u_2(k) - f_2[x_1(k),x_2(k),x_3(k),x_4(k),u_1(k),u_2(k)]\right)\Delta t
\end{bmatrix}
$$

$$
= g(x(k), u(k))
$$

The constraints are

$$
\alpha_1^- \leq u_1(k) \leq \alpha_1^+
$$

$$
\alpha_2^- \leq u_2(k) \leq \alpha_2^+
$$

$$
\beta_1^- \leq x_1(k) \leq \beta_1^+
$$

$$
\beta_2^- \leq x_2(k) \leq \beta_2^+
$$

$$
\beta_3^- \leq x_3(k) \leq \beta_3^+
$$

$$
\beta_4^- \leq x_4(k) \leq \beta_4^+
$$

$$
\beta_5^- \leq u_1(k) - f_1[x_1(k),x_2(k),x_3(k),x_4(k),u_1(k),u_2(k)] \leq \beta_5^+
$$

$$
\beta_6^- \leq u_2(k) - f_2[x_1(k),x_2(k),x_3(k),x_4(k),u_1(k),u_2(k)] \leq \beta_6^+
$$

or, symbolically,

$$
x(k) \ \varepsilon \ X
$$

$$
u(k) \ \varepsilon \ U
$$

The terminal constraint is

$$
x(N) = \begin{bmatrix} x_1(N) \\ x_2(N) \\ x_3(N) \\ x_4(N) \end{bmatrix} = \begin{bmatrix} z_f \\ y_f \\ \dot{z}_f \\ \dot{y}_f \end{bmatrix} = x_f
$$

The initial condition is

$$
x(0) = \begin{bmatrix} x_1(0) \\ x_2(0) \\ x_3(0) \\ x_4(0) \end{bmatrix} = \begin{bmatrix} z_o \\ y_o \\ \dot{z}_o \\ \dot{y}_o \end{bmatrix} = x_o
$$

(c) We define, as usual,

$$
I(x,k) = \min_{u(k),\ldots,u(N-1)} \left\{ \sum_{j=k}^{N-1} L(x(j),u(j))\Delta t \right\}
$$

where

$$
x(k) = x
$$

and where $L(x(k),u(k))$ and Δt are as defined in part (b). We then obtain

$$
I(x,k) = \min_{u} \left\{ L(x,u)\Delta t + I[g(x,u),k+1] \right\}
$$

where $g(x,u)$ is as defined in part (b). The terminal condition is

$$
I(x,N) = \begin{cases} 0, & x = x_f \\ \infty, & x \neq x_f \end{cases},
$$

where x_f is as defined in part (b). At all times the constraints $x \varepsilon X$ and $u \varepsilon U$ are obeyed. The actual value of $I(x,k)$ of most interest to us is $I(x_o,0)$, where x_o is the initial condition defined in part (b). Naturally, all the equations can be expressed in component-by-component form. The basic recursive equation becomes

$$I(x_1,x_2,x_3,x_4,k) =$$

$$\min_{u_1,u_2} \{L(x_1,x_2,x_3,x_4,u_1,u_2)\Delta t +$$

$$I(x_1 + x_3\Delta t, \; x_2 + x_4\Delta t,$$

$$x_3 + \frac{1}{M}[u_1 - f_1(x_1,x_2,x_3,x_4,u_1,u_2)]\Delta t,$$

$$x_4 + \frac{1}{M}[u_2 - f_2(x_1,x_2,x_3,x_4,u_1,u_2)]\Delta t, \; k+1).\}$$

Writing out the constraint equations in component-by-component form is left as an optional exercise.

2.8 Consider the chemical process depicted in Figure 2.4. A unit amount of chemical A at concentration C_{AO} is fed into reactor 1. The cost of any concentration in the allowed range is given by $\phi_{AO}(C_{AO})$.

In reactor 1 the concentration is changed to C_{A1} by applying a temperature T_1. The relation between the resulting concentration, the initial concentration and the temperature is given by $C_{A1} = f_1(C_{AO},T_1)$. The cost of operating the reactor is given by $\phi_{C1}(C_{AO},T_1)$.

The resulting chemical is then fed to a reactor 2 in which the concentration is changed to C_{A2} by adding a catalyst D_2 and by applying a temperature T_2 and a pressure P_2. The new concentration is determined as a function of input concentration, catalyst

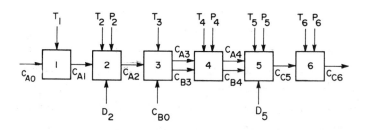

Figure 2.4   Flow Chart of Chemical Process for
Solved Problem 2.7

concentration, temperature, and pressure, according to C_{A2} =
$C_{A2} = f_2(C_{A1}, D_2, T_2, P_2)$. The cost of the catalyst is known to be
$\phi_2(D_2)$, while the cost of operating the reactor is given by
$\phi_{C2}(C_{A1}, D_2, T_2, P_2)$.

The resulting chemical is next mixed in reactor 3 with a unit
amount of another chemical, Chemical B, which has a concentration of
C_{B0}, at a temperature of T_3. This results in the changing of the
concentration of both chemicals to C_{A3} and C_{B3} respectively. The
new concentration of chemical A is given in terms of the input
concentrations of A and B and the temperature according to
$C_{A3} = f_{A3}(C_{A2}, C_{B0}, T_3)$; similarly, the new concentration of Chemical
B becomes $C_{B3} = f_{B3}(C_{A2}, C_{B0}, T_3)$. The cost of Chemical B is known to
be $\phi_{B3}(C_{B0})$, while the cost of operating the reactor is given by
$\phi_{C3}(C_{A2}, C_{B0}, T_3)$.

The resulting mixture of chemicals A and B is then fed into
reactor 4 at a temperature T_4 and a pressure P_4. The concentrations
of the two chemicals are modified to C_{A4} and C_{B4} respectively.
The new concentrations are functions of the input concentrations,
C_{A3} and C_{B3}; the temperature, T_4; and the pressure P_4 according to
the functions $C_{A4} = f_{A4}(C_{A3}, C_{B3}, T_4, P_4)$ and $C_{B4} = f_{B4}(C_{A3}, C_{B3}, T_4, P_4)$.
The cost of operating this reactor is given by $\phi_4(C_{A3}, C_{B3}, T_4, P_4)$.

The resulting mixture then reacts with a catalyst of concentration D_5 at temperature T_5 and pressure P_5 in Reactor 5. The result is a new chemical, Chemical C, at concentration C_{C5}. The concentration of Chemical C is a function of the two input chemical concentrations, the catalyst concentration, the temperature, and the pressure; this function is $C_{C5} = f_5(C_{A4}, C_{B4}, D_5, T_5, P_5)$. The cost of the catalyst at concentration D_5 is $\phi_{D5}(D_5)$, while the cost of operating the reactor is $\phi_{C5}(C_{A4}, C_{B4}, D_5, T_5, P_5)$.

Finally, Chemical C is passed through reactor 6 at a temperature T_6 and pressure P_6. The result is a prescribed amount of Chemical C at concentration C_{C6}; this amount is a function of input concentration, temperature, and pressure given by $C_{C6} = f_6(C_{C5}, T_6, P_6)$. The cost of operating reactor 6 is $\phi_6(C_{C5}, T_6, P_6)$.

The value of the end product, Chemical C at concentration C_{C6}, is $V(C_{C6})$.

Set up dynamic programming recurrence relations to find the input concentrations of Chemicals A and B, C_{A0} and C_{B0}; the catalyst concentrations D_2 and D_5; and the operating temperatures and pressures T_1, T_2, T_3, T_4, T_5, T_6, P_2, P_4, P_5, and P_6, such that the profit of the process is maximized.

This problem illustrates the fact that the number of state variables can change in a process from one stage to the next. We use the normal backward dynamic programming procedure to develop a set of recurrence equations that will allow us to maximize the profit of this process. We begin at the end of the output Chemical C. We call the concentration C_{C6} the state variable $x(6)$. It then follows that the maximum profit at the end of the process is given by

$$I(x(6), 6) = V(x(6))$$

We next define C_{C5}, the concentration at Chemical C at the input
of reactor 6, as the state variable $x(5)$. The variables T_6 and
P_6 are taken to be decision variables $u_1(5)$ and $u_2(5)$ respectively.
We note that $x(6)$ can be expressed in terms of $x(5)$, $u_1(5)$, and
$u_2(5)$ as

$$x(6) = f_6(x(5), u_1(5), u_2(5))$$

The maximum profit that can be achieved over the remainder of the
process starting at the input to reactor 6 with concentration
$x(5) = x$ is thus given by

$$I(x,5) = \max_{u_1,u_2} \quad \{-\phi_6(x,u_1,u_2) + I[f_6(x,u_1,u_2),6]\}$$

We next step back to the inputs to reactor 5. We see that there
are two state variables here, $x_1(4) = C_{A4}$, the input concentration
of Chemical A, and $x_2(4) = C_{B4}$, the input concentration of Chemical
B. The three decision variables are $u_1(4) = D_5$, $u_2(4) = T_5$, and
$u_3(4) = P_5$. We see that $x(5)$ depends on $x_1(4)$, $x_2(4)$, $u_1(4)$,
$u_2(4)$ and $u_3(4)$ according to

$$x(5) = f_5(x_1(4), x_2(4), u_1(4), u_2(4), u_3(4))$$

The maximum profit that can be achieved over the remainder of the
process starting with concentrations $x_1(4) = x_1$ and $x_2(4) = x_2$
is thus given by

$$I(x_1, x_2, 4) = \max_{u_1,u_2,u_3} \quad \{-\phi_{D5}(u_1) - \phi_{C5}(x_1,x_2,u_1,u_2,u_3)$$

$$+ I[f_5(x_1,x_2,u_1,u_2,u_3), 5]\}$$

At the input to reactor 4, we see that there are again two state
variables, $x_1(3) = C_{A3}$ and $x_2(3) = C_{B3}$. The two decision variables
are $u_1(3) = T_4$ and $u_2(3) = P_4$. Using the by-now familiar procedure,
we obtain

$$I(x_1,x_2,3) = \max_{u_1,u_2} \{-\phi_4(x_1,x_2,u_1,u_2)$$

$$+ I[f_{A4}(x_1,x_2,u_1,u_2), f_{B4}(x_1,x_2,u_1,u_2),4]\}$$

We then note that at the input to reactor 3, there is only a single state variable $x(2) = C_{A2}$. The input concentration of Chemical B is a decision variable, $u_1(2) = C_{B0}$. Another decision variable for reactor 3 is the temperature $u_2(2) = T_3$. The two state variables at the output of the reactor are given in terms of the single state variable at the input and the decision variables as

$$x_1(3) = f_{A3}(x(2), u_1(2), u_2(2))$$

and

$$x_2(3) = f_{B3}(x(2), u_1(2), u_2(2))$$

respectively. The recurrence relation for maximum profit starting at the input to reactor 3 with concentration x of Chemical A is then

$$I(x,2) = \max_{u_1,u_2} \{-\phi_{B3}(u_1) - \phi_{C3}(x,u_1,u_2)$$

$$+ I[f_{A3}(x,u_1,u_2), f_{B3}(x,u_1,u_2),3]\}$$

For reactor 2 there is a single state variable at the input, $x(1) = C_{A1}$, and three decision variables, $u_1(1) = D_2$, $u_2(1) = T_2$, and $u_3(1) = T_3$. The recurrence relation becomes

$$I(x,1) = \max_{u_1,u_2,u_3} \{-\phi_2(u_1) - \phi_{C2}(x,u_1,u_2,u_3)$$

$$+ I(f_2[x,u_1,u_2,u_3],2)\}$$

At reactor 1 there is a single state variable, $x(0) = C_{A0}$, and a single decision variable $u(0) = T_1$. The recurrence relation is

$$I(x,0) = \max_{u_1} \{-\phi_1(x,u) + I[f_1(x,u),1]\}$$

Finally, we determine the optimum input concentration of Chemical A
by maximizing the difference between the profit obtained over the
rest of the process by using an input concentration x and cost of
supplying this concentration x. This optimum value, denoted as
$\hat{x}(0)$, is obtained from

$$\hat{x}(0) = \arg \max_{x} \{-\phi_{A0}(x) + I(x,0)\}$$

The complete set of concentrations and operating parameters can
then be obtained by using the optimum decision policy functions
obtained at each stage to trace forward the optimum decision
sequence from $\hat{x}(0)$.

2.9   As an application of the use of functional equations in a
purely mathematical setting, consider the application of dynamic
programming to proving the arithmetic - geometric mean inequality:
Given the numbers

$$a_1, a_2, \ldots, a_N$$

such that

(1) $a_i \geq 0$, $i = 1, 2, \ldots, N$,

and

(2) $\displaystyle\sum_{i=1}^{N} a_i = A$

prove that

$$\left(\frac{1}{N} \sum_{i=1}^{N} a_i\right)^N \geq a_1 \cdot a_2 \cdot a_3 \cdot \ldots \cdot a_N$$

Define

$$I(x,k) = \max_{R(x,k)} \{a_k \cdot a_{k+1} \cdots \cdot a_N\}$$

where $R(x,k)$ is the region $\{a_i : a_i \geq 0, \sum_{i=k}^{N} a_i = x\}$. We can regard this equation as a resource allocation problem of the type consider-ed in Solved Problem 2.5, except that the criterion function is multiplicative. We can immediately establish the recursive equation

$$I(x,k) = \max_{\substack{0 < a_k < A \\ k=0,1,\ldots N-1}} \{a_k I(x - a_k, k+1)\}$$

and determine the value $I(a,1)$; this number represents the maximum value that the product $a_1 a_2 \ldots a_N$ can assume, subject to the constraints $a_i \geq 0$, i-1,2...N and $\sum_{i=1}^{N} a_i = A$. The restriction $0 \leq a_k \leq A$ follows from these constraints. Note that as a starting condition we have

$$I(x,N) = x, \ 0 \leq x \leq A.$$

Let us next show that $I(x,k)$ possesses the structure

$$I(x,k) = \left(\frac{x}{N-k+1}\right)^{N-k+1}, \ k=1,2,\ldots N.$$

Clearly, the result holds for k=N. Assume it holds for k=m+1. Then

$$I(x,m) = \max_{0 \leq a_m \leq A} \left\{ a_m \left[\frac{x-a_m}{N-m}\right]^{N-m} \right\}.$$

Differentiating the term in brackets with respect to a_m and setting the result equal to zero yields

$$\left[\frac{x-a_m}{N-m}\right]^{N-m} - a_m \left[\frac{x-a_m}{N-m}\right]^{N-m-1} = 0$$

The minimizing a_m is thus

$$\hat{a}_m = \frac{x}{N-m+1} \quad .$$

Substituting this value back into the equation for $I(x,m)$ yields

$$I(x,m) = \left(\frac{x}{N-m+1}\right)^{N-m+1}$$

which is the desired result. We apply this result to the quantity $I(A,1)$ to see that

$$I(A,1) = \left(\frac{A}{N}\right)^N = \left[\frac{1}{N}\sum_{i=1}^{N} a_i\right]^N = \max_{R(A,1)} \{a_1 \cdot a_2 \cdot a_3 \cdot \ldots \cdot a_N\}.$$

Therefore

$$a_1 \cdot a_2 \cdot a_3 \cdot \ldots \cdot a_N \leq \left[\frac{1}{N}\sum_{i=1}^{N} a_i\right]^N .$$

Furthermore, if we start from the initial state $x=A$, $k=1$, and recover the optimal decision sequence, we see that at each stage $\hat{a}_k = \frac{A}{N}$; therefore, equality can hold in the above inequality if and only if $a_1 = a_2 = a_3 = \ldots = a_N = \frac{A}{N}$.

2.10 We next consider a resource allocation problem in the realm of boviculture. Assume that we have a herd of cattle with x_i head in age category i, i=0,1,2,...n, where age category i, i=0,1,2,...,n-1, includes all cattle of age i, and age category n includes all cattle of age n or greater. Then at the beginning of each year the following decisions and corresponding returns are available:

(a) cattle may be sent to market, with each head of age category i being worth a_i;

(b) cattle may be kept for breeding, where x_i head of age category i give rise to $b_i x_i$ calves.

Assume that $c_i x_i$ is the fraction of head of age category i which survive to age i+1, $0 \leq c_i \leq 1$.

If the initial herd has d_i head of age category i, i=1,2,...n, determine the breeding and marketing policy which maximizes the total return over N years.

Define the sequence of functions $I(x_0, x_1, \ldots, x_n, k)$ to be the return from the process using an optimal breeding and marketing policy beginning in year k, k=0,1,...N, with x_i cattle of age category i, i=1,2,...n.   Let u_i = the number of head of age category i sent to market in year k, i=0,1,...n.   Then from this allocation we receive an immediate return of

$$\sum_{i=0}^{n} a_i u_i .$$

The new composition of the herd for the following year, year k+1, will be as in Table 2.6.

Table 2.6   Composition of Herd After the k-th Year

i	Cattle of Age i
0	$\sum_{i=0}^{n} b_i (x_i - u_i)$
1	$c_0 (x_0 - u_0)$
2	$c_1 (x_1 - u_1)$
⋮	⋮
n-1	$c_{n-2} (x_{n-2} - u_{n-2})$
$\geq$ n	$c_{n-1} (x_{n-1} - u_{n-1}) + c_n (x_n - u_n)$

The principle of optimality now allows us to assert that

$$I(x_0, x_1, \ldots, x_n, k) = \max_{\substack{0 \leq u_i \leq x_i \\ i=1,2,\ldots,n}} \left\{ \sum_{i=0}^{n} a_i u_i + I\left[\sum_{i=0}^{n} b_i(x_i - u_i), \right. \right.$$

$$c_0(x_0 - u_0), \ c_1(x_1 - u_1), \ldots$$

$$\left. \left. c_{n-1}(x_{n-1} - u_{n-1}) + c_n(x_n - u_n), \ k+1 \right] \right\}$$

Clearly, for the last year (k=N) we sell all cattle and obtain the return

$$\sum_{i=0}^{n} a_i x_i.$$

This gives

$$I(x_0, x_1, \ldots, x_n, N) = \sum_{i=0}^{n} a_i x_i.$$

We then apply the recursive equation to obtain $I(d_0, d_1, \ldots, d_n, 1)$, which is our desired maximum return. The breeding policy can be determined by recovering the optimum decision sequence.

2.11   The next problem we present will illustrate the use of the principle of optimality with a minimax type of criterion function (see Solved Problem 2.1).

    Suppose that members of two different bacterial populations are present and that members of Type I prey upon members of Type II, but not conversely. Furthermore, suppose that neither type is desirable and that we possess a technique (drugs, radiation, etc.) which can be used to destroy members of Type I, but is ineffective against Type II. Determine a policy of control administration such that the maximum of the sum of populations I and II over the time period $0 \leq t \leq T$ is minimized.

If we introduce the notation

$x_1(t)$ = size of population I at time t,

$x_2(t)$ = size of population II at time t,

$u(t)$ = rate of administration of control at time t,

and if the dynamics of the process are described by the differential equations

$$\frac{dx_1}{dt} = g(u(t))x_1 \quad , \qquad x_1(0) = c_1 \quad ,$$

$$\frac{dx_2}{dt} = h(u(t))x_2 - p(u(t))x_1 \quad , \qquad x_2(0) = c_2 \quad ,$$

which hold for x_1, $x_2 \geq 0$, $0 \leq t \leq T$, our objective is to determine the control policy $u(t)$, $0 \leq t \leq T$, such that the functional

$$J = \max_{0 \leq t \leq T} [x_1(t) + x_2(t)]$$

is minimized.

———————————

The crucial point in the solution of this problem is to note that the value of J depends only upon time t and the initial population sizes x_1 and x_2. Thus, let us define the function $I(x_1, x_2, t)$ as

$$I(x_1, x_2, t) = \min_{u(s)} \max_{t \leq s \leq T} [x_1(s) + x_2(s)] \quad .$$

As a result of a decision to use control u, the following events take place:

(1) The initial populations x_1 and x_2 are transformed to the new population levels

$$x_1 \rightarrow x_1 + g(u)x_1 \, \Delta t$$

$$x_2 \rightarrow x_2 + [h(u)x_2 - p(u)x_1] \, \Delta t \quad ,$$

where Δt is the time increment and we again use the Euler integration formula.

(2)   A candidate, $x_1 + x_2$, for the overall maximum cost is
obtained.

We can use the results of Solved Problem 2.1 to observe that the
maximum population sum either occurs immediately, or it occurs
at some later time after an optimal initial control has been applied.
Putting these remarks together with the principle of optimality,
we see that the minimum cost function I satisfies the functional
equation

$$I(x_1,x_2,t) = \max \left\{ x_1 + x_2, \right.$$

$$\left. \min_{u} \left[I(x_1 + g(u)x_1 \Delta t, \; x_2 + [h(u)x_2 - p(u)x_1]\Delta t, \; t + \Delta t \right] \right\}$$

Clearly the terminal condition is

$$I(x_1,x_2,T) = x_1 + x_2.$$

SUPPLEMENTARY PROBLEMS

2.12 Consider the problem of determining the optimal path between two points A and C that arrives at C from a specified direction.

(a)   Let D be an intermediate point along this optimal trajectory (See Figure 2.4). Show that DC is still the optimal path from D to C with respect to this new criterion.

(b)   With the new criterion, is AD an optimal path from A to D?

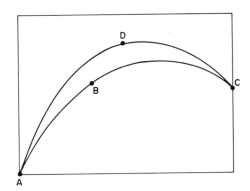

Figure 2.5   Illustration of Trajectories
for Problem 2.12

2.13 Consider the following extensions to the "product of sums" criterion considered in Problem 2.2.

(a)   Develop recursive equations for the "sum of product" criterion given by

$$J = \sum_{i=1}^{N} \left(\prod_{j=1}^{K_i} L_{ij}(u_{ij}) \right)$$

subject to

$$0 \leq u_{ij} \leq \Omega_{ij} \quad , \; j = 1,2,\ldots,K_i, \; i=1,\ldots,N$$

$$0 \leq \sum_{j=1}^{K_i} u_{ij} \leq \Omega_i \; , \; i=1,\ldots,N$$

$$0 \leq \sum_{i=1}^{N} \sum_{j=1}^{K_i} u_{ij} \leq \Omega$$

Again, all functions L_{ij} are assumed to be nonnegative and monotonically nondecreasing functions of their respective arguments u_{ij}.

(b) Establish recursive relations for both the sum of products and product of sums criteria when the constraints take the more general form

$$0 \leq \phi_{ij}(u_{ij}) \leq \Omega_{ij} \quad , \; j=1,2,\ldots,K_i, \; i = 1,\ldots,N$$

$$0 \leq \sum_{j=1}^{K_i} \phi_{ij}(u_{ij}) \leq \Omega_i, \; i = 1,\ldots,N$$

$$0 \leq \sum_{i=1}^{N} \sum_{j=1}^{I_i} \phi_{ij}(u_{ij}) \leq \Omega$$

The functions ϕ_{ij} are assumed to be nonnegative, monotonically nondecreasing functions of their arguments u_{ij}.

(c) For the sum of products and product of sums criteria with the original constraint equations, show how to modify the recursive equations if the L_{ij} are <u>not</u> required to be nonnegative and monotonically nondecreasing.

(d)   Develop recursive equations for the "product of sums of products" criterion given by

$$J = \prod_{\ell=1}^{N} \sum_{i=1}^{N_\ell} \prod_{j=1}^{K_{\ell i}} L_{\ell ij}(u_{\ell ij})$$

and subject to the constraints

$$0 \leq u_{ij} \leq \Omega_{\ell ij} \quad ;$$

$$\sum_{\ell=1}^{M} \sum_{i=1}^{N_\ell} \sum_{j=1}^{K_{\ell i}} u_{\ell ij} \leq \Omega$$

$$J = 1,\ldots,K_{\ell i}, \quad i=1,\ldots,N_\ell, \quad \ell=1,\ldots,M.$$

This problem requires three levels of recursive equations. Characterize and count the number of such equations at each level.

_____

2.14   Repeat the solution for Solved Problem 2.5 when the mass of the vehicle decreases, due to fuel consumption and possibly also staging, according to the known function

$$\frac{dM}{dt} = -f_3(z,y,\dot{z},\dot{y},C_z,C_y,t)$$

The mass at time t_o is known to be M_o. (HINT: Define one additional state variable, $x_5 = M$).

_____

2.15   (Canonical Linear Programming Problem)

Consider the problem of determining the minimum of

$$J = \sum_{i=1}^{N} c_i q_i ,$$

subject to the constraints

$$\sum_{j=1}^{N} a_{ij} q_j \leq P_i, \quad i=1,2,\ldots,M,$$

$$q_i \geq 0, \quad c_i, \quad P_i \text{ real}.$$

Let $I(P_1, P_2, \ldots, P_M, k)$ denote the minimum over the variables $q_k, \ldots, q_M$ of the quantity

$$J_k = \sum_{i=k}^{N} c_i q_i$$

subject to the constraints

$$\sum_{j=k}^{N} a_{ij} q_j \leq P_i, \quad i=k,\ldots,M$$

$$q_i \geq 0, \quad c_i, \quad P_i \text{ real}.$$

Then, show that for all k, $k=1,2,\ldots,N$,

$$I(P_1, P_2, \ldots, P_M, k) = \min_{q_k} \{c_k q_k + I[P_1 - a_{1k}q_k,$$

$$P_2 - a_{2k}q_k, \ldots, P_M - a_{Mk}q_k), k+1]\}$$

where

$$0 \leq q_k \leq \min_{i} \left(\frac{P_i}{a_{ik}} \right).$$

HINT: Observe that as a result of a choice of q_k, the following transformations take place:

(1) a cost $c_k q_k$ is incurred,

(2) the i-th constraint becomes

$$\sum_{i=k}^{N} a_{ij} q_j \leq P_i - a_{ik}q_k, \quad i=1,2,\ldots,M,$$

and

(3)   the number of decisions to be made is reduced from N-k to N-k-1.

———————————

2.16 Given a finite set $\{F_i\}$ of non-negative (each component non-negative) square matrices, let Z_N be the matrix $Q_1 Q_2 \ldots Q_N$, where each Q_i is an F_j. Z_N possesses a characteristic root of largest absolute value. Let r_N be this root. Prove that $\lim_{N \to \infty} (r_N)^{1/N}$ exists.

———————————

2.17 Let us turn to the domain of recreational mathematics and consider the question of map coloring. Numerous mathematical results surround this general class of problem, the most well-known being the famous Four-Color Conjecture. In the problem that follows, this result will be tacitly assumed. This problem, incidently, illustrates nicely the difference between a purely existential versus a constructive method of problem-solving. The famous Four-Color Conjecture, recently proved to be true, simply states that four colors are sufficient to color any planar map, leaving aside the operational question of given a particular map, how one should go about coloring it with the four colors. The results of this problem give a systematic procedure for accomplishing this.

Our problem will concern the region depicted in Figure 2.5. In our map, certain heavy lines have been distinguished by the circled numbers. These lines are called separating boundaries and have the property that the subregions contained between the i-th and (i+1)-st separating boundaries separate the entire region into two parts which have no common boundary. By an overlap we shall mean that two states with a common boundary have the same color.

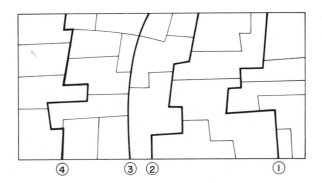

Figure 2.6   Map Problem

The statement of our problem is:

Given that the subregion to the right of separating boundary 1 has been colored according to some coloring scheme, determine a coloring scheme to the left of 1 which will minimize the number of overlaps that occur.

———————————————

REFERENCES

Imbedding and Recurrence Relations: [A-1, [B-8], [B-65], [F-13],
                                      [K-12], [W-1]

Bellman's Principle of Optimality: [B-15], [B-65], [B-83], [H-6],
                                    [L-14], [N-4]

The Optimum Decision Policy: [B-41], [L-14]

Chapter 3

THE BASIC DYNAMIC PROGRAMMING
COMPUTATIONAL PROCEDURE

INTRODUCTION

The purpose of this chapter is to present a detailed exposition of the basic dynamic programming computational procedure.
The computational implications of the principle of optimality are
made explicit here. Also, the application of a digital computer
to solve the basic dynamic programming functional equation is
explained and illustrated in detail.

The chapter begins with a review of the general problem formulation to which the method applies. Next, a brute-force enumeration
procedure is given and its limitations discussed. The basic functional equation derived from the principle of optimality is then reviewed,
and the fundamental difference between dynamic programming and ex-
haustive enumeration is explained.

In the next few sections, the details of the computational
procedure are presented. First, the quantization of the system
variables is discussed. Next, the calculations required to start
the iterations are indicated. Then, the basic calculations for
obtaining the optimal decision are described. Finally, an illustrative example is worked out.

The following section discusses the recovery of an optimal
trajectory from the results of the dynamic programming calculations.
The section after that discusses the interpolation procedures that

are used in both the original calculations and the recovery procedure. Next, alternative implementations of a controller based on
dynamic programming are described. The final section presents in
detail the computational requirements of this procedure.

PROBLEM FORMULATION

As we have seen, the problem to which dynamic programming
applies is the optimization of multistage decision processes.
The essential elements of the problem are the system equations,
which describe the dynamic behavior of the process under consideration; the performance criterion, which evaluates a particular
decision policy; and the constraints, which place restrictions on
the system operations.

The system equations are a set of relations between three
types of variables: the stage variable, the state variables, and
the decision variables.

The stage variable is that quantity which determines the
order in which events occur in the system. This quantity varies
monotonically over the period during which decisions are made;
in many cases it is taken to be time. Thus far, we have assumed
that the stage variable takes on the discrete values k, k=0,1,...,N.
We can also treat the case where the stage variable is continuous
by quantizing it into uniform increments. If we denote the continuous stage variable by t and the range over which t is defined as
$t_o \leq t \leq t_f$, and if we denote the increment size as Δt, then the
quantized stages can be indexed by the discrete sequence k=0,1,...,N,
where the value of t corresponding to k is given by

$$t = t_o + k \, \Delta t \qquad (3.1)$$

and where

$$N\Delta t = t_f - t_o \qquad (3.2)$$

The upper limit of N, can be infinite, but it is generally finite.
More details on the continuous-time case are discussed in Example 3.1.

The state variables are a set of variables that completely
describe the system in the sense that if their values are known
for all k, k=0,1,2,...,N, together with the system inputs (decisions),
then any question about the behavior of the system for this range of
k can be answered. The choice of a set of state variables for a
particular system is not unique, and the determination of a suitable
set is a more or less difficult problem, depending on the extent
and nature of the mathematical model of the system. The choice of
state variables for particular problems has been illustrated in the
previous two chapters, and many more examples will be given later.
For the remainder of this chapter, it will be assumed that for any
given problem a set of n independent state variables can be specified. These n variables are denoted $x_1, x_2, \ldots, x_n$, and they are
written as an n-dimensional vector x, called the state vector.

$$x = \begin{bmatrix} x_1 \\ x_2 \\ \vdots \\ x_n \end{bmatrix} \qquad (3.3)$$

The decision variables are those variables in the process that
can be chosen directly. These variables influence the process
by affecting the state variables in some prescribed fashion. There
are, in general, m decision variables, denoted as $u_1, u_2, \ldots u_m$.
These variables are arranged in an m-dimensional vector, u, called
the decision vector.

$$u = \begin{bmatrix} u_1 \\ u_2 \\ \vdots \\ u_m \end{bmatrix} \qquad (3.4)$$

The system equations describe how the state variables of
stage k+1 are related to the state variables at stage k and the
decision variables at stage k. These equations can be expressed as

$$x_1(k+1) = g_1[x_1(k), x_2(k), \ldots, x_n(k), u_1(k), u_2(k), \ldots, u_m(k), k]$$

$$x_2(k+1) = g_2[x_1(k), x_2(k), \ldots, x_n(k), u_1(k), u_2(k), \ldots, u_m(k), k]$$

$$\vdots \qquad\qquad\qquad\qquad \vdots$$

$$x_n(k+1) = g_n[x_1(k), x_2(k), \ldots, x_n(k), u_1(k), u_2(k), \ldots, u_m(k), k]$$

$$\text{(3.5)}$$

These relations can be written more compactly as

$$x(k+1) = g[x(k), u(k), k] \qquad\qquad (3.6)$$

where g is a n-dimensional vector function.

The criterion function provides an evaluation of a given
decision sequence, u(0), u(1),...,u(N). This criterion is either
an objective function, in which case it is maximized; or else it is a
cost function, in which case it is minimized. Without loss of generality, it will be assumed in this chapter that it is a cost function,
which is to be minimized. The performance criterion usually
depends on each value of u(k) in the decision sequence, and also on
each value of the state vector, x(0), x(1),...,x(N). As discussed
in the last chapter, we shall restrict our formal development to
separable cost function of the form

$$J = \sum_{k=0}^{N} L[x(k), u(k), k] \qquad\qquad (3.7)$$

However, as we shall see later, the computational methods we develop
here can be applied to any criterion function having the Markovian
property.

The constraints place restrictions on the values that the
state variables and decision variables can assume. The state
vector at stage k is constrained to be in the set X(k), which is
a subset of Euclidan n-space R^n. This constraint is expressed
mathematically as

$$x \; \varepsilon \; X(k) \subset R^n \tag{3.8}$$

The decision vector applied at state x, stage k is constrained to
be in the set U(x,k), which is a subset of Euclidan m-space R^m.
This constraint is written

$$u \; \varepsilon \; U(x,k) \subset R^m \tag{3.9}$$

The optimization problem can then be stated as follows:

Given:

    (i)   A system described by Eq. (3.6)

    (ii) Constraints on the state and control as in Eqs.
(3.8) and (3.9)

   (iii) An initial state x(0) = c

Find:

    The decision sequence u(0), u(1),...,u(N) that minimizes
J in Eq. (3.7) while satisfying the constraints.

EXAMPLE

3.1   Consider the continuous-time optimization problem described
by the integral criterion

$$J = \int_{t_o}^{t_f} [L_1(x) + L_2\left(\frac{dx}{dt}\right) + \ldots + L_n\left(\frac{d^{n-1}x}{dt^{n-1}}\right) + L_{n+1}\left(\frac{d^n x}{dt^n}\right)]dt$$

subject to the system differential equation

$$\frac{d^n x}{dt^n} = f(x, \frac{dx}{dt}, \ldots, \frac{d^{n-1}}{dt^{n-1}} , u)$$

(a)   Express the problem in terms of state variables using the
form of Example 1.7 of Chapter 1.

(b)   Discretizing the time interval $[t_o, t_f]$ into uniform increments
Δt, as in Eqs. (3.1) and (3.2), and using the Euler integration
approximation

$$\int_t^{t + \Delta t} \frac{dx}{d\sigma} \, d\sigma \cong \frac{dx}{dt} \, (t) \, \Delta t$$

express the problem in the form of Eqs. (3.6) and (3.7).

(a)   Following the procedure of Example 1.7 of Chapter 1, let u
be the decision variable and define n state variables as

$$x_1 = x$$

$$x_2 = \frac{dx}{dt}$$

$$\vdots$$

$$x_n = \frac{d^{n-1}x}{dt^{n-1}}$$

The criterion then becomes

$$J = \int_{t_o}^{t_f} [L_1(x_1) + L_2(x_2) + \ldots + L_n(x_n)$$

$$+ L_{n+1} [f(x_1, x_2, \ldots, x_n, u)]] \, dt$$

while the system differential equations are

$$\frac{dx_1}{dt} = x_2$$

$$\frac{dx_2}{dt} = x_3$$

$$\vdots$$

$$\frac{dx_{n-1}}{dt} = x_n$$

$$\frac{dx_n}{dt} = f(x_1, x_2, \ldots, x_n, u)$$

(b)   Let $N = \dfrac{t_f - t_o}{\Delta t}$

Then, the integral from t_o to t_f can be split into N integrals
of the form

$$J = \sum_{k=0}^{N-1} \int_{k\Delta t}^{(k+1)\Delta t} [L_1(x_1) + L_2(x_2) + \ldots$$

$$+ L_n(x_n) + L_{n+1} (f(x_1, x_2, \ldots, x_n, u))] \, dt$$

Using the Euler approximation, the desired form of the criterion
becomes

$$J = \sum_{k=0}^{N-1} \{L_1(x_1(k\Delta t)) + L_2(x_2(k\Delta t)) + \ldots + L_n(x_n(k\Delta t))$$

$$+ L_{n+1} (f[x_1(k\Delta t), x_2(k\Delta t), \ldots, x_n(k\Delta t), u(k\Delta t)])\} \Delta t$$

Defining $x_i(k) \overset{\Delta}{=} x_i(k\Delta t)$, $i = 1, 2, \ldots, n$ and $u(k) \overset{\Delta}{=} u(k\Delta t)$, this
equation has the form of Eq. (3.7), where

$$L(x(k), u(k), k) = [L_1(x_1(k)) + L_2(x_2(k)) + \ldots + L_n(x_n(k))$$

$$+ L_{n+1} (f(x_1(k), \ldots, x_n(k), u(k))] \Delta t,$$

$$k = 0, 1, \ldots, N-1$$

$$L(x(k), u(k), N) = 0$$

110   LARSON AND CASTI

To obtain the system difference equation, note that the Euler
approximation yields

$$x(t + \Delta t) - x(t) = \int_{t}^{t+\Delta t} \frac{dx}{d\sigma} d\sigma \cong \frac{dx}{dt}(t) \ \Delta t$$

Setting $t = k\Delta t$, $k=0,1,\ldots,N-1$, applying the above relation to the
n differential equations in part (a) and using the above definitions of $x(k)$ and $u(k)$ yields

$$x_1(k+1) = x_1(k) + x_2(k) \ \Delta t$$

$$x_2(k+1) = x_2(k) + x_3(k) \ \Delta t$$

$$\vdots$$

$$x_{n-1}(k+1) = x_{n-1}(k) + x_n(k) \ \Delta t$$

$$x_n(k+1) = x_n(k) + f(x_1(k), x_2(k),\ldots,x_n(k), u(k)) \ \Delta t$$

$$k=0,1,\ldots,K-1$$

from which one can identify $g(x(k), u(k), k)$ as

$$g_i(x(k), u(k), k) = x_i(k) + x_{i+1}(k)\Delta t \ , \ i=1,2,\ldots,n-1$$

$$g_n(x(k), u(k), k) = x_n(k) + f(x_1(k),x_2(k),\ldots,x_n(k),u(k))\Delta t$$

OPTIMIZATION VIA ENUMERATION OF ADMISSIBLE CONTROL SEQUENCES

Before discussing the dynamic programming approach to
this problem, it is useful to examine a brute-force enumeration
procedure. In this method, the set of admissible decisions, U, is
quantized to a finite number of values. The set U then consists
of the elements

$$U = \{u^{(1)}, u^{(2)},\ldots,u^{(M)}\} \ , \tag{3.10}$$

where both M, the number of admissible decisions, and $u^{(q)}$,
q=1,2,...,M, the actual values of the admissible decisions, can
vary with x and k.

The enumeration procedure then consists of the following steps:
At the given state x(0) every admissible decision u ϵ U is applied.
For each of these decisions, the next state is computed from

$$x(1) = g[x(0), u, 0] \qquad (3.11)$$

If a state x(1) is an admissible state (if x(1) ϵ X(1)), then the cost
associated with this state is evaluated as

$$\Omega[x(1), 1] = L[x(0), u, 0] \qquad (3.12)$$

If the state x(1) is not admissible, then no further consideration
is given to this decision sequence, since it violates the constraints.

This process is continued by applying all admissible decisions
at all of the x(1) ϵ X, finding the resulting states x(2) from

$$x(2) = g[x(1), u, 1] \qquad (3.13)$$

and computing the cost of admissible states x(2) as

$$\Omega[x(2), 2] = L[x(1), u, 1] + \Omega[x(1), 1] \qquad (3.14)$$

In general, when a set of states x(k)ϵX has been defined by
the procedure, a new set of states x(k+1)ϵX is defined by applying
all of u ϵ U at all of the x(k)ϵX, computing the resulting states
from

$$x(k+1) = g[x(k), u, k], \qquad (3.15)$$

and evaluating the cost of admissible states x(k+1) from

$$\Omega[x(k+1), k+1] = L[x(k), u, k] + \Omega[x(k), k] \qquad (3.16)$$

The process continues until N, the final value of k, is reached.

This process traces out all trajectories in state space that
do not violate the constraints and that begin at x(0) and end at
k=N. These trajectories form a tree beginning at x(0), expanding
as k increases. Such a tree is illustrated for a one-dimensional
example in Figure 3.1, where M=2 for all x and k, where X is the
interval $0 \leq x \leq 4$ for all k, and where N = 4. Note that for k = 3
one of the trajectories falls outside of X by taking on too large
a value; while at k=4 two trajectories fall outside of X by being
too large and one by being too small. Thus, it can be seen that
not only is it possible to handle a wide variety of constraints
within this framework, but constraints actually serve a useful
purpose by reducing the number of trajectories that must be
considered.

The minimum cost is evaluated by comparing $\Omega[x(N), N]$ for the
admissible states x(N) ε X and choosing the minimum value. The
optimal decision sequence and the optimal trajectory in state space
are traced out by following back along the tree path that led to
this value of $\Omega[x(N), N]$. This direct procedure always determines
an absolute minimum rather than a relative minimum or maximum and,
if the minimum is not unique, all optimal decision sequences can be
found.

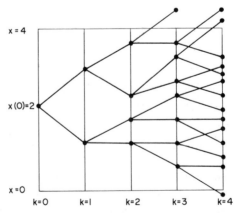

Figure 3.1   Tree Generated by Enumeration

Unfortunately, a straightforward enumeration leads to computational difficulties. Consider a one-dimensional process with M=5 for all x ϵ X and k, with N=20. Then, assuming that none of the generated trajectories violate the constraints, the number of trajectories at k = N = 20 is given by

$$N_T = M^N = 5^{20} \approx 10^{14} \tag{3.17}$$

Storing the 10^{14} costs corresponding to these trajectories, comparing them to find the minimum cost, and tracing back along the tree to find the optimal decision sequence and the optimal trajectory is an infeasible computational task for any existing computer system.

Although the constraints actually reduce the number of trajectories that must be considered, the number of computations still increases exponentially with N. On the basis of the excessive size of the computational requirements for this simple example, it is clear that the procedure is not practical for problems of any degree of complexity.

EXAMPLE

3.2 Consider the problem with system equation

$$x(k+1) = x(k) + u(k),$$

performance criterion

$$J = \sum_{k=0}^{4} [x^2(k) + u^2(k)] + 2.5[x(5) - 2]^2 ,$$

constraints

$$0 \leq x \leq 2 ,$$

$$-1 \leq u \leq 1 ,$$

and initial state

$$x(0) = 2.$$

Quantize the admissible controls to the values

$$U = \{-1,0,1\}$$

Use the enumeration procedure of this section to generate the tree of admissible trajectories. How many of these trajectories are generated at $k=5$? How many state transitions must be considered? How many times must the cost of admissible states be evaluated? What are these numbers if there is no constraint on the state x?

_____

The tree is shown in Figure 3.2. By inspection of this tree, we see that there are 80 admissible trajectories at $k=5$, 147 state transitions, and 128 admissible states (not including $x(0) = 2$) at which the cost must be evaluated. If there were no state constraints, there would be $3^5 = 243$ admissible trajectories, $\sum_{i=1}^{5} 3^i = 363$ state transitions, and 363 admissible states (not including $x(0) = 2$) at which the cost must be evaluated.

THE ITERATIVE FUNCTIONAL EQUATION VS. DIRECT ENUMERATION

In this section we shall examine the iterative functional equation derived in the previous chapter and see if it can be used to suggest improvements in the brute-force enumeration procedure we have just discussed. As a result of this examination, we shall be able to develop a computational procedure that retains all desirable properties of the enumeration procedure, (e.g., determination of an absolute optimum, ease of handling constraints, no assumptions on linearity, etc.) but which has greatly reduced computational requirements.

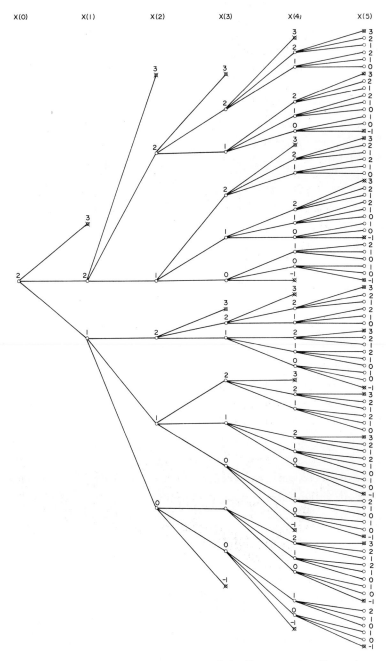

Figure 3.2   Transitions Required in Enumeration Procedure

For the problem formulated earlier in this chapter, we see
from the previous chapter that the following iterative functional
equation is obtained.

$$I(x,k) = \min_{u \varepsilon U} \{L(x,u,k) + I(g(x,u,k), k+1)\} , \qquad (3.18)$$

$$k = 0,1,\ldots,N-1$$

with starting conditions

$$I(x,N) = \min_{u \varepsilon U} \{L(x,u,N)\} \qquad (3.19)$$

The application of these equations in an analytic framework has
been amply illustrated in the previous chapter.

Recall now the principle of optimality enunciated in the
previous chapter. This principle tells us that if an optimal
sequence of states passes through a particular state x at stage k,
then the portion of the sequence from x(k) to the end of the process
must be the optimal sequence from x(k) to the end. Thus, if we work
backward from the end of the process, we do not need to generate a
full decision tree of the type of Figure 3.2 for an intermediate
state x(k), but only to carry the minimum cost function I(x,k) and
the optimum decision û(x,k) associated with this state and stage.

In order to put this concept to use, we need to assume that X,
the set of admissible states, is quantized to some finite number of
values. The set X then contains the elements

$$X = \left\{ x^{(1)}, x^{(2)},\ldots,x^{(J)} \right\} \qquad (3.20)$$

where both J, the number of admissible states, and $x^{(j)}$, $j=1,2,\ldots,J$,
the actual values of these states, can vary with the stage k. The
quantization of U as in Eq. (3.10) is also assumed. The details of
performing this quantization are considered in the next section.

The quantization of U suggests a straightforward method for performing the minimization in Eq. (3.18) - namely, just evaluate the quantity inside the braces for each u ε U and compare these values directly to determine the smallest value. As in the enumeration procedure, two basic calculations are required: first, the cost for the present stage must be evaluated as L(x,u,k); second, the next state g(x,u,k) must be evaluated. The only additional labor required is the evaluation of I(g(x,u,k), k+1), the minimum cost at the next state, and the comparison of the sum L(x,u,k) + I(g(x,u,k), k+1) for each u ε U. As we shall see subsequently, this additional labor is negligible compared to the basic two steps required in the enumeration procedure.

Upon further examination of this procedure, we see that the number of evaluations of L(x,u,k) + I(g(x,y,k), k+1) is once per quantized control for each quantized state for each stage. If the number of quantized controls, quantized states, and stages is fixed at M, J and N respectively, then the number of basic evaluations is M·J·N. Thus, the number of calculations does not grow exponentially with N, as in the case of enumeration, but instead grows linearly. The computational advantages of this latter type of growth versus the former is astronomical.

In order to appreciate what has been accomplished, consider the example discussed in the previous section in conjunction with Eq. (3.17), where M = 5 and N = 20. Let us take a value for J at J = 10, a number consistent with the values of M and K. Then,the total number of evaluations becomes N_c = 5 · 10 · 20 = 1000. If the enumeration procedure is used, there are $5^{20} \approx 10^{14}$ calculations at the last stage above, and a total number of calculations of L(x,u,k) and g(x,u,k) equal to

$$N_n = \sum_{i=1}^{20} 5^i = 5^{21} - 5 \approx 5 \cdot 10^{14}$$

If each such calculation required 10^{-6} seconds on a digital computer,
a reasonable value for current machines even when L(x,u,k) and
g(x,u,k) are complex functions, then in terms of computer time we
are talking about the difference between 10^{-3} seconds and $5 \cdot 10^8$
seconds, i.e., a difference between 1 millisecond and approximately
ten years. For this problem, it is seen that the dynamic programming
approach requires a trivial amount of computational resource, while
the enumeration procedure is totally impractical.

EXAMPLE

3.3   Recall the problem of exercise 3.2. Assume that the set of
admissible states at each stage is X = {0,1,2}. If the dynamic
programming procedure is used, show graphically the state transitions for each of these states for each stage k=0,1,2,3,4. How
many such transitions must be computed? How many times must
L(x,u,k) + I(g(x,u,k), k+1) be computed, including the values at
k=5? Compare these values with the analogous quantities for the
enumeration procedure.

––––––––––––––––––––

The transitions are represented graphically in Figure 3.3.
Note that at k=0 the transitions from x=1 and x=0 have been included,
in addition to those from x=2. While these transitions need not be
calculated in order to solve the original problem, they are included
to show the cost of obtaining a complete feedback decision policy.
The implications of a feedback decision policy are discussed later
in this chapter. Even counting the additional transitions from
x=0 and x=1, it can be seen from the figure that the number of
state transitions that must be considered is $3 \cdot 3 \cdot 5 = 45$.

Note that applying control u = +1 at x=2 results in a next
state x=3, which is not admissible, and that applying control
u = -1 at x=0 results in a new state of x = -1, which is also not
admissible. Since it is not necessary to evaluate

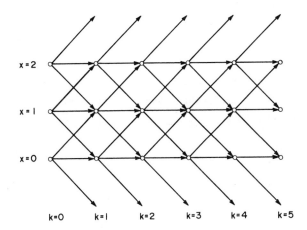

Figure 3.3   Transitions Required by Dynamic Programming
               Procedure

$\{L(x,u,k) + I(g(x,u,k), k+1)\}$ when $g(x,u,k)$ is not an admissible
state, the number of these evaluations actually required is 38, 7
for each of stages k=0,1,2,3,4 and 3 for k=5.   The number of transitions is 147 and the number of evaluations is 128 for the enumeration procedure applied to the same problem.

    The material in this section has outlined the basic procedure
for applying a digital computer to solving multistage decision
problems via dynamic programming.   This section has also shown the
superiority of this approach to a brute-force enumeration procedure,
also suitable for implementation in a computer.   In the remainder
of this chapter we shall discuss the details of all the steps
required to implement the dynamic programming procedure.

CONSTRAINTS AND QUANTIZATION

    One of the most useful properties of the enumeration procedure
discussed in this chapter is that constraints of a very general

nature can be handled.   The constraints actually reduce the computational effort rather than increase it.   Dynamic programming retains
this valuable property.

As in the enumeration procedure, constraints are used to
define X, the set of admissible states.   For example, inequality
constraints of the form

$$\Phi(x,k) \leq 0 \tag{3.21}$$

can be used to restrict the range of the state variables by relations
such as

$$\beta_i^- \leq x_i \leq \beta^+ \quad (i=1,2,\ldots,n) \tag{3.22}$$

The quantities β_i^- and β_i^+ can vary with k.

Constraints are also used to define U, the set of admissible
decisions.   Inequality constraints of the form

$$\Phi(x,u,t) \leq 0 \tag{3.23}$$

can be used to restrict the range of the decision variables

$$\alpha_j^- \leq u_j \leq \alpha_j^+ \quad (j = 1,2,\ldots,q) \tag{3.24}$$

The quantities α_j^- and α_j^+ can vary with both x and k.

Constraints in other forms can generally be manipulated to
reduce to the form of Eqs. (3.23) and (3.24).   Although
the quantities α_j^-, α_j^+, β_i^-, and β_i^+ can vary with x and k, these
variations introduce unnecessary complications into the discussion.
Consequently, for the remainder of this chapter these quantities
will be assumed to be constant.   The extension to the case where
these quantities vary presents no formal difficulties.

For a problem to have physical significance, all state and
control variables must remain finite.   Therefore, if there are no

explicit constraints which limit the range of state and decision variables, then other considerations, such as the range of values which are of interest, are used to establish constraints of the form of Eqs. (3.22) and (3.24).

In order to apply the dynamic programming computational procedure, there must be a finite number of admissible states and admissible decisions. This requirement is usually met by quantizing these variables. Within the range determined by Eq. (3.23), each state variable x_i is quantized with uniform increment Δx_i. These increments could be nonuniform, but again needless notational complications would arise. The extension to the more general case presents no formal difficulties. The quantized values of x_i are thus

$$x_i = \beta_i^- + j_i \Delta x_i \qquad (3.25)$$

where

$$j_i = 0,1,\ldots,J_i$$

$$J_i \Delta x_i = \beta_i^+ - \beta_i^-$$

The set of all x, where each component x_i is quantized according to Eq. (3.25), will from now on be referred to as X, the set of quantized admissible states.

The decision variables can also be quantized according to relations similar to Eq. (3.25). However, in this chapter it is sufficient to assume that there are a finite number of admissible decisions. The set of admissible decisions U, can thus be denoted as

$$U = \{u^{(1)}, u^{(2)},\ldots,u^{(M)}\} \qquad (3.26)$$

where M is the total number of admissible controls

The stage variable has also been quantized and normalized so that k takes on integer values $0,1,2,\ldots,N$, where N is the total number of stages.

STARTING PROCEDURE

In order to use the iterative functional equation, Eq. (3.18), it is necessary to specify a set of boundary conditions. Because the functional equation expresses the minimum cost function at k in terms of the minimum cost function at (k+1), the boundary conditions must be specified at the final stage, N. Formally, the quantity to be determined is I(x,N) for all x ε X. In the computational procedure this quantity is defined by specifying a value of the minimum cost function for every quantized state x ε X. If the cost function is given as J from Eq. (3.7),

$$J = \sum_{k=0}^{N} L[x(k), u(k), k] \qquad (3.27)$$

then I(x,N) can be determined from

$$I(x,N) = \min_{u \in U} \{L[x,u,N]\} \qquad (3.28)$$

If, as is often the case, no decision is made at k=N and hence the cost function at N depends only on the final state, x(N), then I(x,N) can be written directly as

$$I(x,N) = L(x,N) \qquad (3.29)$$

CALCULATION OF OPTIMAL DECISION

Once I(x,N) has been determined for all x ε X, it is possible to compute the optimal decision by iterative application of the functional equation, Eq. (3.18).

Consider a quantized state x ε X at stage (N-1). At this state, each of the admissible deicisons $u^{(m)}$ ε U is applied. For each of these decisions the cost at the current stage can be determined as

$$L^{(m)} = L[x,u^{(m)},N-1] \qquad (m = 1,2,\ldots,M) \qquad (3.30)$$

Next, for each of these decisions the next state at stage N is determined from the system equations, Eq. (3.6)

$$x^{(m)}(N) = g[x, u^{(m)}, N-1] \quad (m=1,2,\ldots,M) \qquad (3.31)$$

The next step is to compute the minimum cost at stage N for each of the states $x^{(m)}$. However, in general a particular state $x^{(m)}$ will not lie on one of the quantized states $x \in X$ at which the optimal cost $I(x,N)$ is defined. In fact, it may lie outside of the range of admissible states determined by Eq. (3.22). In the latter case the decision is rejected as a candidate for the optimal decision for this state and stage.

If a next state $x^{(m)}$ does fall within the range of allowable states, but not on a quantized value, then it is necessary to use some type of interpolation procedure to compute the minimum cost function at these points. In general, the interpolation procedure consists of using a low-order polynomial in the n state variables to approximate the minimum cost function in small regions of state space. The coefficients of the polynomial are determined in terms of the known values of the minimum cost function at quantized states $x \in X$. The determination of the coefficients is made according to some criterion, such as least-squares fit. Computational procedures for calculating the coefficients are relatively simple and well-known. Interpolation formulas are discussed in more detail in a later section of this chapter.

Assume, then, that the values of the minimum cost at the states $x^{(m)}$ can be expressed as a function of the values of the optimal cost at quantized states $x \in X$.

$$I[x^{(m)},N] = P[x^{(m)}, N, I(x,N)], \quad (\text{all } x \in X) \qquad (3.32)$$

The total cost of applying decision $u^{(m)}$ at state x, stage (N-1), can then be written as

$$F_1^{(m)} = L[x,u^{(m)}, N-1] + I[x^{(m)},N] \qquad (3.33)$$

This quantity is exactly the function which is to be minimized by choice of $u^{(m)}$ in the functional equation, Eq. (3.18). The minimization can be achieved by simply comparing the M quantities. According to the functional equation, the minimum value will be the minimum cost at state x , stage (N-1).

$$I[x,N-1] = \min_{u^{(m)} \varepsilon U} \{L[x,u^{(m)}, N-1] + I[x^{(m)},N]\} \qquad (3.34)$$

The optimal decision at this state and stage, $\hat{u}[x,N-1]$, is the control $u^{(m)}$ for which the minimum in Eq. (3.34) is actually taken on.

This procedure is repeated at each quantized state $x\varepsilon X$ at stage (N-1). When this has been done, $I(x,N-1)$ and $\hat{u}(x,N-1)$ are known for all $x \varepsilon X$. It is now possible to compute $I(x,N-2)$ and $\hat{u}(x,N-2)$ for all $x \varepsilon X$ based on knowledge of $I(x,N-1)$.

The general iterative procedure continues this process. Suppose that $I(x,k+1)$ is known for all $x \varepsilon X$. Then $I(x,k)$ and $\hat{u}(x,k)$ are computed for all $x \varepsilon X$ from

$$I(x,k) = \min_{u^{(m)} \varepsilon U} \{L[x,u^{(m)},k] + I[x^{(m)}, k+1]\} \qquad (3.35)$$

where $x^{(m)}$ is determined from

$$x^{(m)} = g[x,u^{(m)},k] \qquad (3.36)$$

and where $I[x^{(m)}, k+1]$ is computed by interpolation on the known values $I[x, k+1]$ for all $x \varepsilon X$:

$$I[x^{(m)},k+1] = P[x^{(m)}, k+1, I(x, k+1)], \text{ (all } x \varepsilon X) \qquad (3.37)$$

The optimal decision $\hat{u}(x,k)$ is the decision for which Eq. (3.35) takes on the minimum. The iterative procedure begins by computing

$\hat{u}(x,N-1)$ and $I(x,N-1)$ from the given boundary conditions $I(x,N)$, and it continues until $\hat{u}(x,0)$ and $I(x,0)$ have been computed.

The procedure can be understood most easily in terms of a simple example. Such an example is worked out in the next section. Also, a flow chart illustrating the general procedure is presented in the following section.

AN ILLUSTRATIVE EXAMPLE

In this section a simple optimization problem will be solved by dynamic programming in order to illustrate the basic computational procedure. The system equation is taken to be

$$x(k+1) = x(k) + u(k) \qquad (3.38)$$

where $x(k)$ = scalar state variable at stage k.
$\quad\;\; u(k)$ = scalar decision variable at stage k.
The performance criterion, which is to be minimized, is

$$J = \sum_{k=0}^{9} [x^2(k) + u^2(k)] + 2.5[x(10) - 2]^2. \qquad (3.39)$$

The state variable is constrained to lie in the interval

$$0 \leq x \leq 8, \qquad (3.40)$$

while the decision variable is bounded by

$$-2 \leq u \leq 2. \qquad (3.41)$$

The final state of the system at k = 10, is constrained to fall in the interval

$$0 \leq x(10) \leq 2. \qquad (3.42)$$

The state variable is quantized in uniform increments of one, i.e., $\Delta x=1$. Thus, $X=\{0,1,2,3,4,5,6,7,8\}$ is the set of admissible states.

The decision variable is also quantized in uniform increments
of 1, so that the set of admissible decisions is

$$U = \{-2, -1, 0, 1, 2\} \tag{3.43}$$

With these quantizations there is no need to perform any interpolation, which makes the calculations easier to follow.

The grid of quantized values of x and k at which computations
are to be made is shown in Figure 3.4. When the computational procedure is about to begin, all final states with x(10)>2 are forbidden
by the constraint of Eq. (3.42). This is denoted by placing x's at
all these states. A small circle is placed at x(10) = 0, x(10) = 1,
and x(10) = 2 to indicate that a minimum cost can be computed for
those points. From Eq. (3.39) it can be seen that these minimum
costs are

$$I(0,10) = 10 \tag{3.44}$$
$$I(1,10) = 2.5$$
$$I(2,10) = 0$$

These values are placed in Figure 3.4 to the right of the grid
points to which they correspond.

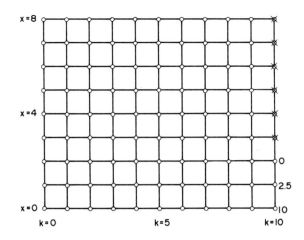

Figure 3.4   Initial Conditions for Dynamic Programming
Computational Procedure

With the initial values of minimum cost established, it is
possible to apply iteratively the functional equation based on the
principle of optimality, Eq. (3.18). Making the identifications

$$L[x(k), u(k), k] = x^2(k) + u^2(k) \quad , \quad (k=0,1,2,\ldots,9) \qquad (3.45)$$

and

$$g[x(k), u(k), k] = x(k) + u(k), \quad (k=0,1,2,\ldots,9) \qquad (3.46)$$

the iterative relation becomes

$$I(x,k) = \min_{u \in U} \{x^2 + u^2 + I(x + u, \; k+1)\}, (k=0,1,2,\ldots,9) \qquad (3.47)\cdot$$

The set of admissible decisions U is given in Eq. (3.43). The
values $I(x,10)$ in Eq. (3.44) are used as starting conditions. If
an admissible decision results in a next state which is not admissible (the next state either fails to satisfy Eq. (3.40) or else it
has an x placed on it), then that decision is not considered further.
If all admissible decisions are rejected on this basis, then an x
is placed at such a point, and it is considered an inadmissible state
for this stage.

Equation (3.47) is first applied at k=9, then at k=8, then at
k=7, etc., until k=0 is reached. Each time an admissible optimal
decision is found, a small circle is drawn at that grid point, the
optimal decision is written to the right and below the circle, and
the minimum cost is written to the right and above it. These
numbers constitute the dynamic programming solution to the problem.

The computations at a given state x for k=9 take place as
follows: each decision u ε U is applied and the next state is computed for each decision. If the next state is admissible, the minimum
cost of that state is found from Eq. (3.44). The cost $[x^2(9) + u^2(9)]$,
over the next stage is computed and the sum of the two costs is
stored. If the next state is not admissible, nothing is stored.
From among the stored costs the minimum is picked. This cost
becomes the minimum cost at state x, stage k, while the optimal
decision is the decision corresponding to the minimum cost.

For $x(9) = 0$ the computations can be summarized in tabular form as follows:

Table 3.1   Computations at x=0, k=9

u	$g(x,u,k)$	$I(g,k+1)$	$L(x,u,k)$	Total Cost
-2	-2	x	x	x
-1	-1	x	x	x
0	0	10	$0^2 + 0^2 = 0$	10
1	1	2.5	$0^2 + 1^2 = 1$	3.5
2	2	0	$0^2 + 2^2 = 4$	4

The x's in the table indicate that the next state is not admissible. Comparing the total costs for admissible next states, it is seen that the minimum cost is 3.5, which corresponds to u=1. Note that this minimum utilizes neither the minimum "immediate cost," $L(x,u,k)$ which in this case corresponds to u=0, nor the minimum "next-state cost", $I(g,k+1)$, which in this case corresponds to u=2. It is the combination of these two quantities which must be minimized.

For $x(9) = 1$, the computations can be abbreviated in the following form

Table 3.2   Computations at x=1, k=9

u	$g(x,u,k)$	$I(g,k+1)$	$L(x,u,k)$	Total Cost
-2	-1	x	x	x
-1	0	10	$1^2 + (-1)^2 = 2$	12
0	1	2.5	$1^2 + 0^2 = 1$	3.5
1	2	0	$1^2 + 1^2 = 2$	2
2	3	x	x	x

The minimum cost is 2, again corresponding to u=1.

For x(9) = 2,

Table 3.3   Computations at x=2, k=9

u	g(x,u,k)	I(g,k+1)	L(x,u,k)	Total Cost
-2	0	10	$2^2 + (-2)^2 = 8$	18
-1	1	2.5	$2^2 + (-1)^2 = 5$	7.5
0	2	0	$2^2 + 0^2 = 4$	4
1	3	x	x	x
2	4	x	x	x

The minimum cost is 4, corresponding to u=0.

For x(9) = 3,

Table 3.4   Computations at x=3, k=9

u	g(x,u,k)	I(g,k+1)	L(x,u,k)	Total Cost
-2	1	2.5	13	15.5
-1	2	0	10	10
0	3	x	x	x
1	4	x	x	x
2	5	x	x	x

The minimum cost is 10, corresponding to u=-1.

For x(9) = 4,

Table 3.5   Computations at x=4, k=9

u	g(x,u,k)	I(g,k+1)	L(x,u,k)	Total Cost
-2	2	0	20	20
-1	3	x	x	x
0	4	x	x	x
1	5	x	x	x
2	6	x	x	x

The minimum cost is 20, corresponding to u=-2, the only decision which results in an admissible next state.

For x(9)>4, there is no decision which results in an admissible next state. Consequently, these states must be regarded as inadmissible. This completes the computations at k=9. The results are summarized in Figure 3.5.

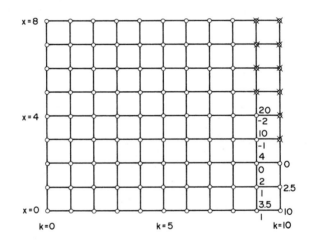

Figure 3.5   Results After Computations at k=9

The computations at k=8 are then performed using Eq. (3.35). The values of minimum cost at k=9 are used as I(g,k+1) in this relation. The computations take place exactly as at k=9. For example, at x(8) = 0,

Table 3.6   Computations at x=0, k=8

u	g(x,u,k)	I(g,k+1)	L(x,u,k)	Total Cost
-2	-2	x	x	x
-1	-1	x	x	x
0	0	3.5	0	3.5
1	1	2	1	3
2	2	7	4	11

The minimum cost is 3, corresponding to u=1.

EXAMPLE

3.4 Construct tables for all other admissible states at k=8.

The appropriate tables are shown below.

Table 3.7   Computations at x=1, k=8

u	g(x,u,k)	I(g,k+1)	L(x,u,k)	Total Cost
-2	-1	x	x	x
-1	0	3.5	2	5.5
0	1	2	1	3
1	2	4	2	6
2	3	10	5	15

The minimum cost is 3, corresponding to u=0.

Table 3.8   Computations at x=2, k=8

u	g(x,u,k)	I(g,k+1)	L(x,u,k)	Total Cost
-2	0	3.5	8	11.5
-1	1	2	5	7
0	2	4	4	8
1	3	10	5	15
2	4	20	8	28

The minimum cost is 7, corresponding to u=-1.

Table 3.9   Computations at x=3, k=8

u	g(x,u,k)	I(g,k+1)	L(x,u,k)	Total Cost
-2	1	2	13	15
-1	2	4	10	14
0	3	10	9	19
1	4	20	10	30
2	5	x	x	x

The minimum cost is 14, corresponding to u=-1.

Table 3.10   Computations at x=4, k=8

u	g(x,u,k)	I(g,k+1)	L(x,u,k)	Total Cost
-2	2	4	20	24
-1	3	10	17	27
0	4	20	16	36
1	5	x	x	x
2	6	x	x	x

The minimum cost is 24, corresponding to u=-2.

Table 3.11   Computations at x=5, k=8

u	g(x,u,k)	I(g,k+1)	L(x,u,k)	Total Cost
-2	3	10	29	39
-1	4	20	26	46
0	5	x	x	x
1	6	x	x	x
2	7	x	x	x

The minimum cost is 39, corresponding to u=-2.

Table 3.12   Computations at x=6, k=8

u	g(x,u,k)	I(g,k+1)	L(x,u,k)	Total Cost
-2	4	20	40	60
-1	5	x	x	x
0	6	x	x	x
1	7	x	x	x
2	8	x	x	x

The minimum cost is 60, corresponding to u=-2.

There are no admissible next states for any admissible decisions from x=7 and x=8.

After the computations at k=8 have been completed, the resulting minimum costs are used in the computations at k=7. This procedure, in which optimal decision and minimum cost at stage k are
computed using minimum costs at (k+1), continues until k=0 is
reached.

EXAMPLE

3.5 Complete the solution and fill in the results on the grid.

The completed grid is shown in the figure below.

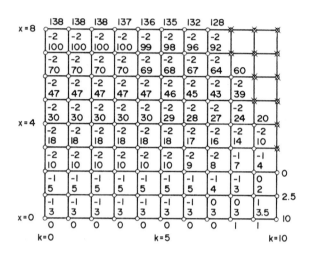

Figure 3.6   Complete Results of Dynamic Programming Computational Procedures for Illustrative Example

EXAMPLE

3.6 An important consideration in implementing the dynamic programming solution is the variation with the stage k of the optimal decision policy $\hat{u}(x,k)$ and the minimum cost function $I(x,k)$.

(a) In this illustrative example, are there any states for which the optimal decision or the minimum cost function does not vary with k?

(b) Do either the optimal decision function or the minimum cost function reach a value at some stage k for which the function does not change at values of k less than this value? If so, for what value of k does this occur?

(c)   If the state transition function $g(x,u,k)$ and the single-
stage cost function $L(x,u,k)$ take on the same values as in Eqs.
(3.45) and (3.46) for stage $k = -1$, what would the minimum cost
function and optimal decision policy be for this stage?

(a)   There is no state for which the optimum decision policy or
minimum cost function is constant for all k. However, for the states
x=4, x=5, x=6, x=7, and x=8, the optimal decision is always u=-2
at those stages where the state is admissible. The optimum decision
is undefined at a stage for which the state is inadmissible.

(b)   Yes, in both cases. The optimum decision function is fixed
for k≤6 while the minimum cost function is fixed for k≤2. The
functions are said to have reached their steady-state values when
this occurs.

(c)   As will be discussed in Chapter 4, for a system with a time-
in-variant state transition function and time-invariant single-
stage cost function, once the optimum decision function and minimum cost function reach the steady-state condition, they do not
change any further. Thus, these functions have the same values at
k=-1 as they do at k=0. The reader can verify this directly.

REVIEW OF THE DYNAMIC PROGRAMMING COMPUTATIONAL PROCEDURE AND A
FLOW CHART FOR A COMPUTER PROGRAM TO IMPLEMENT IT

A flow chart to implement the procedure of the previous
section on a digital computer is shown in Figure 3.7. First, state,
control, and stage variables are quantized. Constraints are used
to restrict the admissible values of state and decision variables.
The stage variable is indexed as k, k=0,1,2,...,N. The set of
quantized admissible states, X, is indexed as $x^{(j)}$, j=1,2,...,J.
The set of admissible controls, U, is indexed as $u^{(m)}$, m=1,2,...,M.

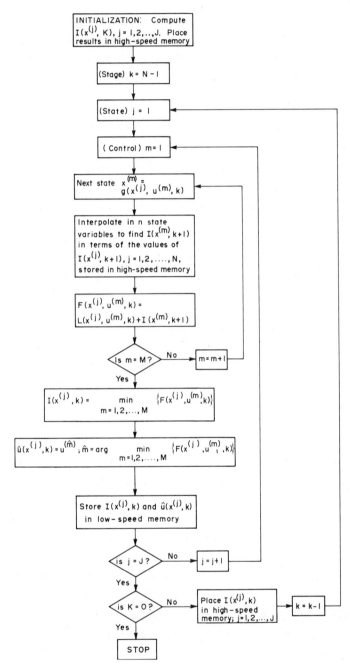

Figure 3.7   Flow Chart of the Conventional Dynamic
              Programming Computational Procedure

The initial conditions for the functional equation are determined as $I[x^{(j)},N]$, $j=1,2,\ldots,J$. The functional equation is used to compute the minimum cost function $I(x,k)$ as

$$I[x^{(j)},k] = \min_{u^{(m)}} \{L[x^{(j)},u^{(m)},k] + I[x^{(m)},k+1]\}, m=1,2,\ldots,M \qquad (3.48)$$

where

$$x^{(m)} = g(x^{(j)},u^{(m)},k) \qquad (3.49)$$

and where $I[x^{(m)},k+1]$ is determined by using the interpolation formula and the known values $I[x^{(j)}, k+1]$. The optimal control is determined as

$$\hat{u}[x^{(j)},k] = u^{(\hat{m})} \qquad (3.50)$$

where $\hat{m}$ is the index of m for which the quantity in brackets in Eq. (3.48) is minimized. As illustrated in the flow chart, the procedure consists of determining $I[x^{(j)},k]$ and $\hat{u}[x^{(j)},k]$ in terms of $I[x^{(j)}, k+1]$, for every quantized state $x^{(j)}$, $j=1,2,\ldots,J$, beginning at stage (N-1) and continuing until stage 0 is reached.

RECOVERY OF AN OPTIMAL TRAJECTORY

The dynamic programming solution is a specification of $\hat{u}(x,k)$ and $I(x,k)$ for all quantized $x \in X$ for $k=0,1,2,\ldots,N$. However, the original problem was to find the optimum sequence of decisions starting from the given $x(0)$. This sequence can easily be determined from the dynamic programming solution. The first decision in the sequence is evaluated as

$$\hat{u}(0) = \hat{u}[x(0),0] \qquad (3.51)$$

The next state along the sequence is then found by applying the system difference equation to obtain

$$\hat{x}(1) = g[x(0), \hat{u}(0), 0] \qquad (3.52)$$

This state may or may not be a quantized state. If it is, then
the next decision in the optimum sequence is evaluated directly as

$$\hat{u}(1) = \hat{u}[\hat{x}(1),1] \tag{3.53}$$

If it is not a quantized state, then values of $\hat{u}(x,1)$ for a number
of quantized states, $x \in X$, must be found and a suitable interpolation formula used. These interpolations follow a procedure similar
to that indicated for the minimum cost function in Eq. (3.32) and
discussed further in a later section.

The recovery procedure, in which the next state is computed on
the basis of the present state and the optimal decision function
evaluated at the present state utilizing the function $\hat{u}(x,k)$,
continues until the complete decision sequence, $\hat{u}(0)$, $\hat{u}(1),\ldots,\hat{u}(N)$,
and optimal trajectory $\hat{x}(0)$, $\hat{x}(1),\ldots,\hat{x}(N)$, have been obtained.

For the solution shown in Figure 3.7,the optimum decision
sequence and optimal trajectory from $x(0) = 8$ are shown in Table
3.13. The decisions $\hat{u}(\hat{x},k)$, $k=0,1,2,\ldots,10$, are read directly from
the figure (no interpolations are required), while the states along
the trajectory are obtained from

$$\hat{x}(k+1) = \hat{x}(k) + \hat{u}(k) \quad (k=0,1,2,\ldots,9) \tag{3.54}$$

Table 3.13   Optimum Control Sequence and
              Optimal Trajectory from x(0)=8

k	$\hat{x}$	$\hat{u}$	$L(\hat{x},\hat{u},k)$
0	8	-2	68
1	6	-2	40
2	4	-2	20
3	2	-1	5
4	1	-1	2
5	0	0	0
6	0	0	0
7	0	0	0
8	0	1	1
9	1	1	2
10	2	-	0

$$\text{Minimum Cost} = \sum_{k=0}^{10} L(\hat{x},\hat{u},k) = 138$$

PROPERTIES OF THE DYNAMIC PROGRAMMING COMPUTATIONAL PROCEDURE

It can be seen at this point that the dynamic programming
computational procedure has a number of very desirable properties.
In the first place, it is not necessary to make any assumptions
about the analytic properties of the system difference equation,
$x(k+1) = g(x(k), u(k), k)$, or of the single-stage cost function,
$L(x(k), u(k), k)$. These functions are not required to be linear,
quadratic, differentiable, continuous or even expressible in terms
of well-known functions. All that is required is the existence of
a rule for determining values of these functions at quantized values
of the state x, the decision u, and the stage k and a procedure for
interpolating between quantized values. The procedure thus can
accommodate highly nonlinear systems, as well as system equations
that implement logical operations and/or experimentally tabulated
phenomena. The solved problems at the end of the chapter will
further illustrate this point.

A second desirable property of the dynamic programming procedure is the ease with which it handles constraints. As was noted a few sections ago, constraints of an extremely general type can be implemented by merely decreasing the size of X(k), the set of admissible states at stage k, and U(x,k), the set of admissible decisions at state x and stage k. Thus, constraints present only minor difficulties in implementation, and actually serve a useful purpose by decreasing the number of alternative states and/or decisions that must be considered.

A third desirable property of the procedure is that it always determines an absolute minimum, not a relative minimum, maximum, or even worse, a stationary point. This property is a result of the fact that all quantized admissible states are considered at each stage and that for each state all quantized admissible decisions are considered. Thus, within the accuracy of the quantization, a true global optimum is always obtained.

Still another favorable property of the procedure is its inherent simplicity. The only calculations required are stepping forward of the system difference equation, looking up and/or interpolating the minimum cost function at the next stage, and comparing scalar quantities. This simplicity not only makes computer implementation of the procedure quite straightforward, but it also allows workers of diverse technical backgrounds to thoroughly understand the method and to feel confident in its application

Of course, all of these properties were also present in the brute-force enumeration procedure discussed early in the chapter. However, all the desirable properties have been retained in dynamic programming, while the computational requirements have been drastically reduced.

The procedure also possesses another very important property lacking in the enumeration procedure. This property is related to the fact that solutions are obtained for an entire family of

problems, not just for a single problem. This occurs because we
obtain the minimum cost and optimum decision at every admissible
quantized state and stage.

Bellman refers to the process of solving an entire family of
problems in order to solve a given, more complex, problem as invariant
imbedding. A most important consequence of this approach is that a
feedback control or decision policy solution is obtained. This simply
means that the optimal decision is specified as a function of both
state and stage, not just as a function of the stage, as in the case
in open-loop control. The importance of this type of solution is
that in the case of deviations from the original optimal trajectory
as, for example, might occur if an uncontrolled input were applied
to the system or if an incorrect decision was inadvertently implemented, a truly optimal decision can be found for the remaining
stages; if the solution were a function of stage only, the remainder
of the decision sequence would be incorrect unless further computations were performed.

EXAMPLE

3.7   Consider the illustrative example discussed in detail a few
sections ago. Assume that the system state is $x(6) = 1$. The optimum
decision sequence and optimal trajectory are shown in Table 3.14.
Now assume that the decision $u(6) = 0$ is applied, rather than the
optimal decision $u(6) = -1$, with the result that the system state
becomes $x(7) = 1$ rather than $x(7) = 0$.

(a)   Referring to the complete solution shown in Figure 3.6, what is
the optimal trajectory from the state $x(7) = 1$? What is the corresponding total cost?

(b)   What trajectory is followed if the original decision sequence
from Table 3.14 is followed from the state $x(7) = 1$? What is the
corresponding cost?

Table 3.14   Optimum Decision Sequence and
Optimal Trajectory from x(6)=1

k	$\hat{x}$	$\hat{u}$	$L(\hat{x},\hat{u},k)$
6	1	-1	2
7	0	0	0
8	0	1	1
9	1	1	2
10	2	-	-

$$\text{Total Cost} = \sum_{k=6}^{10} L(\hat{x},\hat{u},k) = 5$$

(c)   What trajectory is followed if the system is returned from
x(7) = 1 as fast as possible to the original trajectory specified
in Table 3.14?   What is the corresponding cost?

———————————

(a)   From examination of Figure 3.6, the optimal decision sequence
and optimal trajectory from $\hat{x}(7)$ = 1 is as shown in Table 3.15.

(b)   If the decision sequence for $\hat{u}(7)$, $\hat{u}(8)$, and $\hat{u}(9)$ from Table
3.14 is applied at state x(7) = 1 instead of the true optimum
sequence shown in Table 3.15, then the trajectory in Table 3.16
will be followed.   Note that the resulting final state, x(10), is
inadmissible; thus, in this case, the open-loop decision sequence
would actually be inadmissible.

(c)   If a control u(7) were applied at state x(7) = 1 such that
x(8) took on the value zero and thus was back on the open-loop
optimal trajectory (a task that would require additional computation)
the resulting trajectory from $\hat{x}(7)$ = 1 is as shown in Table 3.17.
The cost of the trajectory is 5, rather than 4 as found in Table
3.15 using the feedback solution.

Table 3.15   Optimum Decision Sequence and
              Optimal Trajectory from x(7)=1

k	$\hat{x}$	$\hat{u}$	$L(\hat{x},\hat{u},k)$
7	1	0	1
8	1	0	1
9	1	1	2
10	2	–	0

$$\text{Total Cost} = \sum_{k=7}^{10} L(\hat{x},\hat{u},k) = 4$$

Table 3.16   Result of Applying Open-Loop Optimum
              Decision Sequence as Shown in Table
              3.14 for State x(7)=1

k	$\hat{x}$	$\hat{u}$	$L(\hat{x},\hat{u},k)$
7	1	0	1
8	1	1	2
9	2	1	5
10	3	–	X

$$\text{Total Cost} = \sum_{k=7}^{10} L(\hat{x},\hat{u},k) = X$$

Table 3.17   Result of Applying Decision Sequence
              that Returns to Original Optimal Tra-
              jectory from State x(7) = 1

k	$\hat{x}$	$\hat{u}$	$L(\hat{x},\hat{u},k)$
7	1	-1	2
8	0	1	1
9	1	1	2
10	2	–	0

$$\text{Total Cost} = \sum_{k=7}^{10} L(\hat{x},\hat{u},k) = 5$$

INTERPOLATION PROCEDURES

In an earlier section it was noted that an interpolation of $I(x,k)$ in the n state variables is required in the basic dynamic programming computational procedure. The need for a similar interpolation of $\hat{u}(x,k)$ in the utilization of results was also mentioned. The purpose of this section is to describe briefly how these interpolations are performed.

A standard numerical analysis technique for interpolation is multi-dimensional polynomial approximation, and a considerable number of methods and formulas can be found in the literature. For practical reasons, however, only the simplest of these formulas have been used extensively in dynamic programming applications.

The most sophisticated methods involve using a high-order polynomial over a large number of data points. However, such methods generally require excessive computing time. Not only is a considerable amount of time necessary for calculating the coefficients of the polynomial, but also much time is used in evaluating the high-order polynomial every time an interpolation is required.

A second approach is to cover many data points but to use a low-order polynomial. In this case, the polynomial must be fitted to the data according to some criterion, such as least-squares error. Unfortunately, such an approach usually results in a considerable loss of accuracy. Since the polynomial does not pass through all the data points, interpolated values can differ considerably from the known values at near-by data points. Furthermore, considerable computing time is required to calculate these coefficients on the basis of the data points.

A third approach is to use a high-order polynomial, but to cover only a small number of data points. However, experience has

indicated that the optimal cost function $I(x,k)$ tends to vary
fairly smoothly in a small region, and the use of a high-order
polynomial tends to introduce unrealistically large variations
between data points. Limited success has been obtained in approxi-
mating $\hat{u}(x,k)$ by a high-order polynomial over even relatively small
regions.

The only remaining alternative, use of a low-order polynomial
over a small region, is the most widely practiced procedure in
dynamic programming. The simplest approach is to approximate the
function by the value at the nearest quantized state. For a one-
dimensional example with x quantized in uniform increments Δx from
x=0 to x=aΔx, the equations for interpolating $I(x,k)$ are

$$I(x,k) = I(a\Delta x, \ k), \qquad (3.55)$$

for

$$(a-\tfrac{1}{2}) \ \Delta x < x \leq (a+\tfrac{1}{2})\Delta x$$

A similar equation can be written for $\hat{u}$.

The next simplest approach is to use a multi-dimensional
linear approximation between data points, where the linear approximation is exact at the data points. The equations for the one-
dimensional example specified previously are

$$I(x,k) = I(a\Delta x,k) + \frac{I[(a+1)\Delta x,k] - I(a\Delta x,k)}{\Delta x} (x-a\Delta x) \qquad (3.56)$$

for

$$a\Delta x \leq x \leq (a+1)\Delta x$$

Again, a similar formula exists for $\hat{u}(x,k)$.

The highest-order formula that receives extensive use in
practice is the multi-dimensional quadratic approximation.

EXAMPLE

3.8 For the one-dimensional case where exact fit is made at the
data points, find the formula for quadratic interpolation.

_____

The form of the solution is assumed to be

$$I(x,k) = A + B(x-a\Delta x) + C(x-a\Delta x)^2 \tag{3.57}$$

where A, B, and C are to be determined by exactly fitting the values
$I(a\Delta x,k)$, $I((a+1)\Delta x,k)$ and $I((a+2)\Delta x,k)$. Fitting first at $x=a\Delta x$,
we find A immediately as

$$I(a\Delta x,k) = A \tag{3.58}$$

Next, fitting at $x=(a+1)\Delta x$,

$$I((a+1)\Delta x,k) = A + B\Delta x + C(\Delta x)^2 \tag{3.59}$$

Finally, fitting $x=(a+2)\Delta x$,

$$I((a+2)\Delta x,k) = A + 2B\Delta x + 4C(\Delta x)^2 \tag{3.60}$$

Solving the latter two equations for C yields

$$C = \frac{1}{(\Delta x)^2} \left\{ \frac{1}{2} I((a+2)\Delta x,k) - I((a+1)\Delta x,k) + \frac{1}{2} I(a\Delta x,k) \right\} \tag{3.61}$$

Finally, we solve for B as

$$B = \frac{1}{\Delta x} \left\{ -\frac{1}{2} I((a+2)\Delta x,k) + 2I((a+1)\Delta x,k) \right.$$

$$\left. - \frac{3}{2} I(a\Delta x,k) \right\} \tag{3.62}$$

Substituting into Eq. (3.57), we obtain the desired relation as

$$I(x,k) = I(a\Delta x,k)$$

$$+ \; [- \frac{1}{2} I((a+2)\Delta x,k) + 2I((a+1)\Delta x,k)$$

$$- \frac{3}{2} I(a\Delta x,k)] \frac{(x-a\Delta x)}{\Delta x}$$

$$+ \; [\frac{1}{2} I((a+2)\Delta x,k) - I((a+1)\Delta x,k)$$

$$+ \frac{1}{2} I(a\Delta x,k)] \; \frac{(x-a\Delta x)^2}{(\Delta x)^2} \tag{3.63}$$

where the range of validity is from $a\Delta x \leq x \leq (a+2)\Delta x$. If the approximation is to be used only over the interval of length Δx where it is most accurate, then the range should be restricted to $(a+\frac{1}{2})\Delta x \leq x \leq (a+3/2)\Delta x$.

In other cases, particularly multi-dimensional problems, more data points are used than can be fit exactly by the interpolation formula. It is then necessary to determine the coefficients in the formula by least-squares fit techniques; the coefficients are selected by minimizing the sum of the squares of the deviations of the interpolation formula from the actual values at data points.

Because these interpolation formulas are not exact, errors are introduced both into the calculation of $I(x,k)$ and $\hat{u}(x,k)$ from the iterative functional equation and into the recovery of the optimal trajectory from values of $\hat{u}(x,k)$. In determining which formulas to use in a specific application, it is necessary to consider not only the instantaneous effects of these errors, but also the propagation of these errors into subsequent calculations. Bellman has studied the propagation of these errors in the computation of the iterative functional equation; he has shown that, under relatively modest assumptions, the effects of the errors made at a given stage diminish as the number of stages computed increases. This property of the iterative functional equation is extremely

important, and it adds to the desirability of dynamic programming
as a computational tool. By using results on the stability of
difference equations, it is possible to obtain similar results
about the errors in trajectory recovery.

IMPLEMENTATION OF DYNAMIC PROGRAMMING SOLUTION

As discussed in a previous section, the dynamic programming
solution, $\hat{u}(x,k)$, leads to a feedback control or decision policy
configuration. One method of implementing this solution is to
simply store all the values of $\hat{u}(x,k)$ in memory, monitor the
state and stage of the system, and look up the appropriate value
of $\hat{u}(x,k)$ as required. This type of implementation is attractive
because the dynamic programming calculations can be done off-line,
and the only operation that needs to be done during the decision
interval is retrieval of the appropriate optimal decision. The
system configuration is as shown in Figure 3.8.

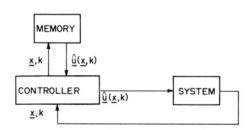

Figure 3.8   A Controller Based on Retrieving the
               Results of the Dynamic Programming
               Computation from Memory

An alternative scheme is to carry out the dynamic programming
computations in real-time. In this case, the state and stage are
again monitored, but the entire optimum decision sequence is re-
computed by an on-line computer; this re-computation may be
required as often as once every stage, but if deviations from the
computed trajectory are small the number of re-computations can be

much less. Generally, a nominal trajectory based on either the
last-computed trajectory, the results of pre-computations, or
operating experience is used to decrease the size of the sets X and
U over which computations are made. This type of approach is considered in more detail in Volume 2 of this series. This implementation is pictured in Figure 3.9; note that a feedback control solution
is still obtained.

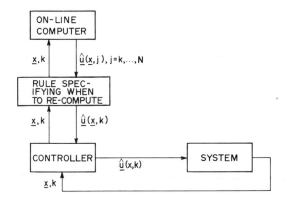

Figure 3.9   A Controller Based on Use of an On-Line
              Computer

Still another method of utilizing the dynamic programming
results is as a guide in developing a sub-optimal decision policy
realization. One approach is to search for simple functions that
closely approximate $\hat{u}(x,k)$. Generally, this involves analyzing the
numerical values of $\hat{u}(x,k)$ to see what analytical functions closely
approximate them and using least-squares fit procedures to find the
best combinations of these functions. In any case, the minimum
cost function for the original problem, $I(x,k)$, serves to evaluate
the performance loss of any sub-optimal controller generated from
the optimal controller.

COMPUTATIONAL REQUIREMENTS

Before concluding this chapter, let us examine the computational
requirements of the procedure we have been discussing. The numerous
desirable properties cited previously for the procedure provide us
with a great incentive for applying it to numerous problems in many
diverse fields; however, as we shall see shortly, the computational
requirements of the procedure grow rapidly with the number of state
and decision variables, a phenomenon Bellman refers to as the "curse
of dimensionality". At this point in the development of the digital
computer, it is extremely difficult to apply the straightforward
procedure to problems having more than six state variables and/or
six decision variables. Nevertheless, continuing developments in
the computer design field, including both hardware developments and
new concepts in computer architecture, show great promise for extend-
ing the applicability of the method; as an example of what can be
achieved in the latter area, the reader is referred to previous work
by the authors and their colleagues on the subject of implementation
of dynamic programming on parallel architecture computers. Furthermore, many researchers have previously taken note of both the great
desirability of the method and its "curse of dimensionality", and
they have developed numerous modifications of the procedure to
mitigate the latter while retaining the former; the reader interested
in applying dynamic programming to high-dimensional problems is
urged to consult Volume 2 of this series, where many of the most
promising of these methods are reviewed and numerous others are
referenced.

The most commonly encountered barrier to the use of dynamic
programming is the high-speed memory requirements. This requirement
refers to the number of locations in the high-speed access memory
(generally, core memory) which must be available during the computations. In addition to the locations needed for the program, the compiler, and other special functions (a number of locations which may be

large fraction of the total storage available), the dynamic programming procedure requires that sufficient data be stored to specify
$I(x,k)$ for all $x \in X$ at a single value of k. In general, this is
done by storing one value of $I(x,k)$ for every quantized $x \in X$ and
by using a simple interpolation formula as discussed earlier. The
number of locations required is then

$$N_h = \prod_{i=1}^{n} N_i \qquad (3.64)$$

where N_i = number of quantized values of ith state variable
      n = total number of state variables.

If current values of minimum cost are also to be retained in the
high-speed memory, the number of locations required is increased to
$2N_h$. If computations are taking place at stage k, the N_h values
$I(x,k+1)$ must be stored for interpolation purposes. However, as
the values $I(x,k)$ are generated, they also must be stored so that
computations using these results can be performed at stage (k-1).
The total number of locations required to store both $I(x,k-1)$ and
$I(x,k)$ is thus $2N_h$.

It is clear that the number N_h can be extremely large for
moderate-sized problems. For example, if there are 6 state variables (n=6) and if there are 10 quantization levels in each
variable (N_i=10, i=1,2,...,6) then

$$N_h = 10^6 \qquad (3.65)$$

The number 10^6 is at the limit of the total high-speed storage capacity fo the largest existing computers. (As many as 2 or 3 x 10^6
storage locations are available with some computers as of this
writing; however, machines with this capacity are extremely costly.)
Of course, a significant fraction of this number may be reserved
for the program, the compiler and other special functions. Thus this
modest problem saturates the high-speed memory of most existing
computers.

A second storage problem arises in retaining the results of
the computation. If there are N_h quantized states at each stage
and if there are N stages, then the number of values of $\hat{u}(x,k)$
and of $I(x,k)$ which are computed is N_c where

$$N_c = N_h \cdot N. \tag{3.66}$$

Although this number can be extremely large, currently available
low-speed memory deivces, such as magnetic tape and bulk disk
storage, are capable of storing this much information reliably and
at a reasonable cost. In the example where $N_h = 10^6$, which
saturates the high-speed memory of most available computers, a
value of N = 50 implies that

$$N_c = 50 \cdot 10^6 \tag{3.67}$$

This number, while rather large, is feasible for currently available
systems.

The computing time requirement is related to N_c, the number of
points at which $I(x,k)$ and $\hat{u}(x,k)$ are computed, as well as to N_d,
the number of discrete values of u that must be tried at each such
point. The number N_d is determined from

$$N_d = \prod_{j=1}^{m} M_j \tag{3.68}$$

where M_j = number of quantized values of jth decision variable,
      m = number of decision variables

If at a given value of x and k it takes Δt_c seconds to compute the
quantity $\{L(x,u,k) + I(g(x,u,k), k+1)\}$ and to compare it with
other values, then the total computation time becomes

$$T_c = N_c \cdot N_d \cdot \Delta t_c \tag{3.69}$$

In the example we have been discussing, if there are two decision
variables, each quantized to 10 levels (M_j = 10, j=1,2), then

$$N_d = 10^2 \tag{3.70}$$

For a value of Δt_c = 10^{-6} sec, a reasonable value for the fastest
computers currently available, the computing requirement becomes

$$T_c = 50 \cdot 10^6 \cdot 10^2 \cdot 10^{-6} = 5,000 \text{ seconds} \tag{3.71}$$

This time, about 83 minutes, is justifiable for many problems of
current interest.

It is useful to compare the above computing time with that
required for enumeration. Since there are 100 possible values of
$\hat{u}(x,k)$ at each x and k, and N=50, the number of computation is

$$N_T \approx 100^{50} = 10^{100} \tag{3.72}$$

If the amount of time for each computation is taken to be $\Delta t_T = 10^{-6}$ sec,
which is consistent with the value of Δt_c = 10^{-6} sec, the total
computing time for direct enumeration is

$$T_T = N_T \cdot \Delta t_T = 10^{100} \cdot 10^{-6} = 10^{94} \text{ seconds}, \tag{3.73}$$

which is approximately $3 \cdot 10^{86}$ years, a time about $2 \cdot 10^{90}$ times greater
than the value of T_c for the same problem. Similar calculations
verify that in all but the very simplest examples dynamic programming
requires orders of magnitude less computing time than direct enumeration.

Despite the great advantage over enumeration, the preceding
results indicate that in problems having 6 or more state variables,
the computational requirements may exceed the capacity of even the
largest and fastest computers available. If the quantization of
the variables is required to be finer, i.e., if the values of N_i,
M_j and N are larger, then problems are encountered even with

three or four state variables. Fortunately, however, the advanced
computational procedures to be discussed in Volume 2 allow us to
extend the basic method to problems of substantially higher dimension.

SUMMARY

This chapter has presented a thorough discussion of the basic
dynamic programming computational procedure. Every step of the
procedure has been explained in detail and illustrated in the context of a simple but meaningful example. These explanations have
allowed us to see the strong points of the procedure: the ability
to handle system equations and performance criteria of a very general
type (nonlinear, nonquadratic, discontinuous, nonanalytic, dependent
on tabular data, etc.); the ease of dealing with constraints; the
determination of an absolute optimum within quantization accuracy;
the inherent simplicity, both in comprehension and implementation;
the feedback control or complete decision policy solution. The
explanations have also clearly spelled out the computational requirements of the procedure and allowed us to evaluate the feasibility
of applying the procedure to specific problems; more sophisticated
procedures that allow us to retain the desirable properties of
dynamic programming while attacking high-dimensional problems beyond
the scope of this procedure are discussed in Volume 2.

Before leaving this chapter, the reader should attempt to
acquire one more item: enough experience in solving problems with
the procedure to be able to apply it with confidence. This degree
of appreciation of the method can only be obtained by working out
examples. The solved problems section of this chapter provides an
opportunity to analyze how the method can be applied in a variety
of application areas, and several of these problems will allow the
reader to see how a relatively unstructured problem formulation is
reduced to a point where the method can be applied. Perhaps one of
the most exciting aspects of dynamic programming is its ability to

provide both an extremely powerful conceptual point of view for structuring approaches to difficult and important problems and a practical computational tool for obtaining numerical results in such problems. It is hoped that the worked problems and supplementary exercises give the reader some appreciation of both these facets of dynamic programming.

SOLVED PROBLEMS

3.1 Consider the problem with performance criterion

$$J = \sum_{k=0}^{4} [x^2(k) + u^2(k)] + 2.5(x(5) - 2)^2,$$

system equation

$$x(k+1) = x(k) + u(k),$$

and constraints

$$0 \leq x \leq 2,$$

$$-1 \leq u \leq 1,$$

Use uniform quantization increments $\Delta x = 1$ and $\Delta u = 1$.

(a) Develop the complete dynamic programming solution

(b) Trace out the solution from $x(0) = 2$

(c) Trace out the solution from $x(0) = 0$

(d) Trace out the solution from $x(2) = 1$

(a) For the given quantization increments, the set of admissible states is $x = \{0,1,2\}$ and the set of admissible decisions is $U = \{-1,0,+1\}$. Following the standard procedure, the solution grid is as shown in Figure 3.10, where the number to the right and above the grid point is $\hat{u}(x,k)$ and the number to the right and below the grid point is $I(x,k)$. Note that at x=1, k=1, there are two values of u, u=0 and u=-1, which both attain the minimum cost $I(1,1) = 5$. As a result of this "tie" for the optimal decision, there can be more than one solution to the optimization problem corresponding to certain initial states. This non-uniqueness of solution causes no difficulties in implementing the procedure. Either all solutions can be retained or one of the solutions can be selected on an arbitrary basis.

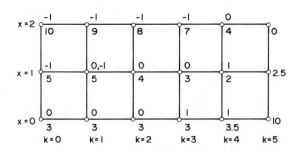

Figure 3.10   Dynamic Programming Solution

(b)   For initial state x(0) = 2 there are two solutions. They are
shown in the two tables below.

Table 3.18   First Optimum Solution for
x(0) = 2

k	$\hat{x}$	$\hat{u}$	$L(\hat{x},\hat{u},k)$
0	2	-1	5
1	1	-1	2
2	0	0	0
3	0	1	1
4	1	1	2
5	2	–	0

Total Cost = 10

Table 3.19   Second Optimum Solution for x(0)=2

k	$\hat{x}$	$\hat{u}$	$L(\hat{x},\hat{u},k)$
0	2	-1	5
1	1	0	1
2	1	0	1
3	1	0	1
4	1	1	2
5	2	–	0

Total Cost = 10

The two trajectories in state-stage space corresponding to these
two solutions are shown in Figure 3.11.

(c) For the initial state x(0) = 0, the unique solution is given in
Table 3.20.

Table 3.20   Optimum Solution for x(0) = 0

k	$\hat{x}$	$\hat{u}$	$L(\hat{x},\hat{u},k)$
0	0	0	0
1	0	0	0
2	0	0	0
3	0	1	1
4	1	1	2
5	2	–	0

Total Cost = 3

The trajectory corresponding to this solution is shown in Figure
3.12.

(d) For the initial state x(2) = 1, the unique solution is given in
the table shown below.

Table 3.20   Optimum Solution for x(2)=1

k	$\hat{x}$	$\hat{u}$	$L(\hat{x},\hat{u},k)$
2	1	0	1
3	1	0	1
4	1	1	2
5	2	–	0

Total Cost = 4

The trajectory corresponding to this solution is shown in Figure
3.13.

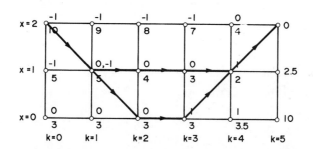

Figure 3.11   Trajectories for the Two Solutions
from x(0) = 2

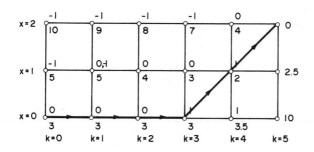

Figure 3.12   Trajectory for Solution from x(0) = 0

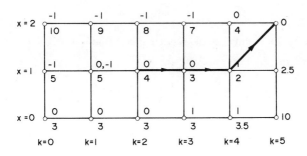

Figure 3.13   Trajectory for Solution from x(2) = 1

3.2 Consider the problem with performance criterion

$$J = \sum_{k=0}^{4} [x^2(k) + u^2(k)]$$

system equation

$$x(k+1) = x(k) + u(k)$$

and constraints

$$0 \le x \le 2$$
$$-1 \le u \le 1$$

Use uniform quantization increments $\Delta x=1$ and $\Delta u=1$. Note this problem is the same as the previous problem except for deleting the terminal cost function $2.5 [x(5) - 2]^2$.

(a) Develop the complete dynamic programming solution

(b) Trace out the solution from $x(0) = 2$

(c) Trace out the solution from $x(0) = 0$

(d) Trace out the solution from $x(2) = 1$

(e) Explain any differences between the solutions to this problem and the previous problem .

_____

(a) For the given quantization increments the set of admissible states is $X = \{0,1,2\}$ and the set of admissible decisions is $U = \{-1,0,1\}$. Following the standard procedure, the solution grid is as shown in Figure 3.14, where $\hat{u}(x,k)$ and $I(x,k)$ are placed on the figure as in Figure 3.10.

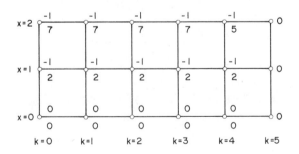

Figure 3.14   Dynamic Programming Solution

(b)   For the initial state x(0) = 2, the solution is shown in Table
3.22.

Table 3.22   Optimum Solution for x(0) = 2

k	$\hat{x}$	$\hat{u}$	$L(\hat{x},\hat{u},k)$
0	2	-1	5
1	1	-1	2
2	0	0	0
3	0	0	0
4	0	0	0
5	0	-	0

Total Cost = 7

The trajectory corresponding to this solution is shown in Figure
3.15.

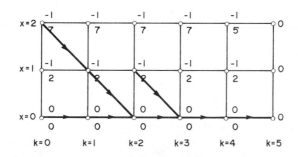

Figure 3.15   Trajectories for
x(0) = 2, x(0) = 0, and x(2) = 1

(c)   For the initial state x(0) = 0, the solution is shown in Table
3.23.   The trajectory corresponding to this solution is shown in
Figure 3.15.

(d)   For the initial state x(2) = 1, the solution is shown in Table
3.24.   The trajectory corresponding to this solution in shown in
Figure 3.15.

Table 3.23   Optimum Solution for x(0) = 0

k	$\hat{x}$	$\hat{u}$	$L(\hat{x},\hat{u},k)$
0	0	0	0
1	0	0	0
2	0	0	0
3	0	0	0
4	0	0	0
5	0	–	0

Total Cost = 0

Table 3.24   Optimum Solution for x(2) = 1

k	$\hat{x}$	$\hat{u}$	$L(\hat{x},\hat{u},k)$
2	1	-1	2
3	0	0	0
4	0	0	0
5	0	–	0

Total Cost = 2

(e)   The differences between these results and those of Example 3.1 are due to leaving out the terminal cost function.   In this case, there is no incentive to return the state to x=2 at the final stage, k=5.   For stages k=1 and 2 the decision policies are essentially the same, but for k $\geq$ 3 the absence of the terminal penalty changes the strategy.

3.3   Consider the problem with performance criterion

$$J = \sum_{k=0}^{9} [x^2(k) + 5u^2(k)] + 2.5\,[x(10) - 2]^2,$$

system equation

$$x(k+1) = x(k) + u(k),$$

and constraints

$$0 \leq x(k) \leq 8 , \quad k=0,1,\ldots,9$$
$$0 \leq x(10) \leq 2,$$
$$-2 \leq u(k) \leq 2. \quad k=0,1,\ldots,9$$

Use uniform quantization increments $\Delta x=1$ and $\Delta u=1$. Note that this problem is the same as the illustrative example worked in Chapter 3, except that the weighting on u is 5 rather than 1.

(a) Find the complete dynamic programming solution to this problem.

(b) Trace out the solution from $x(0) = 8$. Referring to Figure 3.17, compare this solution to that for $x(0) = 8$ in the illustrative example. Explain any differences.

(c) Trace out the solution from $x(3) = 4$. Referring to Figure 3.17, compare this solution to that for $x(3) = 4$ in the illustrative example. Explain any differences.

(a) For the given quantization increments, the set of admissible states is $X = \{0,1,2,3,4,5,6,7,8\}$ and the set of admissible decisions is $U = \{-2,-1,0,1,2\}$. Following the standard procedure, the complete solution grid is as shown in Figure 3.16, where $\hat{u}(x,k)$ and $I(x,k)$ are placed on the figure as in Figure 3.16.

(b) For initial state $x(0) = 8$, the solution is given in Table 3.25, and the corresponding trajectory is marked on Figure 3.16.

The corresponding solution to the illustrative example is given in Table 3.26, and the corresponding trajectory is marked on the complete solution grid in Figure 3.17.

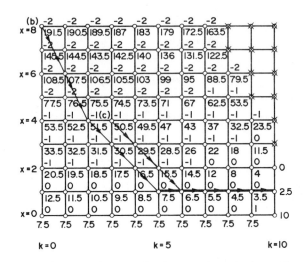

Figure 3.16   Complete Solution Grid Marked to Show
Trajectories from x(0)=8 and x(3)=4

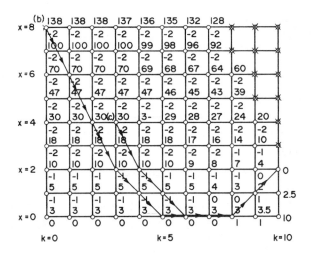

Figure 3.17   Complete Solution Grid for Illustrative
Example Marked to Show Trajectories
x(0)=8 and x(3)=4

Table 3.25   Optimum Solution for x(0) = 8

k	$\hat{x}$	$\hat{u}$	$L(\hat{x},\hat{u},k)$
0	8	-2	84
1	6	-2	56
2	4	-1	21
3	3	-1	14
4	2	-1	9
5	1	0	1
6	1	0	1
7	1	0	1
8	1	0	1
9	1	0	1
10	1	-	2.5

Total Cost = 191.5

Table 3.26   Optimum Solution for x(0)=8 in
Illustrative Example

k	$\hat{x}$	$\hat{u}$	$L(\hat{x},\hat{u},k)$
0	8	-2	68
1	6	-2	40
2	4	-2	20
3	2	-1	5
4	1	-1	2
5	0	0	0
6	0	0	0
7	0	0	0
8	0	1	1
9	1	1	2
10	2	-	0

Total Cost = 138

In Example 3.3 there is a larger penalty on the square of the decision variable, and hence the magnitude of the values of u are smaller. Instead of driving the state to x=0 as fast as possible and then returning it to x=2 at k=10, the state is only driven to x=1 by k=5 and held there until k=10. The larger penalty thus prevents the system from being driven to the state for which the penalty or the square of the state is smallest (x=0 for k < 10 and x=2 for k=10).

(c) For the initial state x(3)=4, the solution is given in Table 3.27.

The corresponding solution to the illustrative example is shown in Table 2.38, and the corresponding trajectory is shown on the complete solution grid in Figure 3.17.

The comments given in part (b) regarding the differences in these solutions apply here as well.

Table 3.27   Optimum Solution for x(3)=4

k	$\hat{x}$	$\hat{u}$	$L(\hat{x},\hat{u},k)$
3	4	-1	21
4	3	-1	14
5	2	-1	9
6	1	0	1
7	1	0	1
8	1	0	1
9	1	0	1
10	1	-	2.5

Total Cost = 50.5

Table 3.28   Optimum Solution for x(3)=4 in
Illustrative Example

k	$\hat{x}$	$\hat{u}$	$L(\hat{x},\hat{u},k)$
3	4	-2	20
4	2	-1	5
5	1	-1	2
6	0	0	0
7	0	0	0
8	0	1	1
9	1	1	2
10	2	-	0

Total Cost = 30

3.4   The examples considered so far have involved relatively simple
linear and quadratic functions in the performance criterion and
system equations. This example illustrates that the same computational procedure extends in a straightforward manner to more complicated functions. Consider the performance criterion

$$J = \sum_{k=0}^{4} (2 + u(k))3^{-x(k)} + |x(5) - 1|,$$

system equations

$$x(k+1) = x(k) + [x - 2x(k) + \frac{5}{4} x^2(k) - \frac{1}{4} x^3(k)]u(k),$$

and constraints

$$0 \leq x \leq 3$$
$$-1 \leq u \leq 1$$

Use a uniform quantization $\Delta x = 1$ and $\Delta u = 1$ and employ linear
interpolation between values at the two nearest quantized states for
any interpolations required.

(a)   Develop the complete dynamic programming solution.

(b)   Trace out the solution from x(0) = 2.

(a)   For the given quantization increments, the set of admissible
states is X = {0,1,2,3}, and the set of admissible decisions
is U = {-1,0,1}. In order to facilitate the analysis, it is useful
to calculate the state transition function and the single-stage
cost function for each quantized state as a function of the decision
variable u. Since both of these functions are time-invariant, it
saves time to compute them once and store them. The results of
these calculations appear in Tables 3.29 and 3.30.

Table 3.29   State Transition Function

x(k)	x(k+1)
0	2u(k)
1	1 + u(k)
2	2 + u(k)
3	3 + 0.5 u(k)

Table 3.30   Single-Stage Cost Function

x(k)	L(x(k),u(k),k)
0	2 + u(k)
1	0.36788 [2 + u(k)]
2	0.13534 [2 + u(k)]
3	0.04979 [2 + u(k))]

Note that at x=3 the decision u=-1 results in a next state of 2.5.
Since this state is not one of the quantized states in X, a linear
interpolation will be necessary to obtain the minimum cost of the
next state.

The calculations begin by specifying I(x,5) at the quantized
states as shown in Figure 3.18. To illustrate the interpolation
procedure, consider the three controls applied at x=3, k=4 and

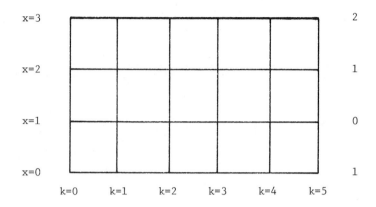

Figure 3.18   Boundary Conditions

refer to Table 3.31. For u=+1, the next state is x=3.5, which
violates the state constraint. For u=0, the next state is x=3.
The minimum cost at the next state is I(3,5) = 2, and the single-
stage cost is L(3,0,4)=0.09958, resulting in a total cost of
2.09958. For u=-1, the next state is x=2.5, half-way between x=2
and x=3. A linear interpolation between the cost I(2,5)=1 and
I(3,5)=2 yields a cost I(2.5,5)=1.5. The single-stage cost is
L(3,-1,1) = 0.04979, resulting in a total cost of 1.54979. This
latter cost is clearly the minimum value.

Table 3.31   Calculation at x=3, k=4

u	g(x,u,k)	I(g,k+1)	L(x,u,k)	Total Cost
1	3.5	x	x	x
0	3	2	0.09958	2.09958
-1	2.5	1.5	0.04979	1.54979

The complete solution is shown in Figure 3.19.

0	-1	-1	-1	-1	
0.83720	0.73762	0.69898	0.89236	0.54979	2
1	0	0	0	-1	
1.14363	0.94735	0.67669	0.40601	0.13534	1
1	1	1	1	0	
2.05099	1.78033	1.50965	1.23898	0.73576	0
1	1	1	1	0	
3.94735	3.67669	3.40601	3.13534	3.00000	1

Figure 3.19   Complete Solution Using Linear
Interpolation

(b) For the initial state $x(0) = 2$, the solution is shown in Table 3.32. Note that this solution requires interpolation to obtain the optimal decision for $k \geq 2$.

Table 3.32   Optimum Solution for $x(0) = 2$

k	x	u	L(x,u,k)
0	2	1	0.40601
1	3	-1	0.04979
2	2.50	-0.50	0.12313
3	2.05	-0105	0.24792
4	2.00	-1.00	0.13534
5	1.00	-	0.00000

Total Cost = 0.96219

The optimum decision for $x(0) = 2$ is u=1, which results in a next state of $x(1) = 3$ and a single-stage cost of 0.40601.

The optimum decision for x(1) = 3 is u=-1, which results in a next state of x(2) = 2.50 and a single-stage cost of 0.04979.

From this state on we require interpolations. Linear interpolation of the decision policy function between $\hat{u}(2,2) = 0$ and $\hat{u}(3,2) = 1$ yields $\hat{u}(2.5,2) = 0.50$. These values of state and decisions can now be substituted into the original system equations and single-stage cost functions. This substitution shows that the next state is x(3) = 2.05 and that the single-stage cost is 0.12313.

Interpolation between $\hat{u}(2,3) = 0$ and $\hat{u}(3,3) = -1$ yields the optimum decision $\hat{u}(2.05,3) = -0.05$. Substitution into the system equations and single-stage cost function shows that the next state is x(4) = 2.00 and the single-stage cost is 0.24792.

Next we obtain directly that the corresponding optimal decision is $\hat{u}(4,4) = -1$, the single-stage cost at k=4 is 0.13534, and the final state is x(5) = 0.00.

Finally, we note that the terminal penalty corresponding to this final state is 0.00.

It is instructive to note that the cost along this trajectory is 0.96219 rather than the 1.14363 computed in the dynamic programming solution . This discrepancy shows that for highly nonlinear system equations and cost functions, such as those used here, it is sometimes necessary either to use a finer quantization interval and/ or higher order interpolation to obtain precise results.

3.5   Repeat the solution to problem 3.4 using quadratic interpolation in the minimum cost function and linear interpolation in the optimum decision policy.

———————————

(a)   The calculations proceed as in Example 3.4 except for the quadratic interpolation at x = 3. The complete results are shown in Figure 3.20.

To illustrate the difference made by the quadratic interpolations, consider the calculations at x=3, k=4. First, we attempt to obtain a formula for $I(x,5)$ by using an exact fit for the quadratic formula at x=1, x=2, and x=3. If we denote the interpolation formula by

$$I(x,5) = a + bx + cx^2,$$

then we see that we need to solve the equations

$$a + b + c = 0$$
$$a + 2b + 4c = 1$$
$$a + 3b + 9c = 2$$

The solution is

$$a = -1$$
$$b = 1$$
$$c = 0$$

The formula is thus,

$$I(x,5) = -1 + x.$$

In this case, the quadratic interpolation formula is the same as the linear interpolation formula and the same result as shown in Table 3.31 is obtained.

0	-1	-1	-1	-1	2
0.63317	0.53359	0.43999	0.64150	1.54979	
1	0	0	0	-1	1
0.93960	0.94735	0.67669	0.40601	0.13534	
1	1	1	1	0	0
2.05099	1.78033	1.50965	1.23898	0.73576	
1	1	1	1	0	1
3.94735	3.67669	3.40601	3.13534	3.00000	

Figure 3.20   Quadratic Interpolation

For x=3, k=3, the situation is more complicated. Again, we
attempt to obtain an interpolation formula of the form

$$I(x,4) = a + bx + cx^2$$

by obtaining an exact fit at x=1, x=2 and x=3. The equations in
this case are

$$a + b + c = 0.73676$$
$$a + 2b + 4c = 0.13534$$
$$a + 3b + 9c = 0.54979$$

The solution is

$$a = 3.35106$$
$$b = -3.62274$$
$$c = 1.00744$$

Using the formula to evaluate I(2.5,4), we obtain

$$I(2.5,4) = 0.59171$$

The complete calculations at x=3, k=3 are then as shown in Table
3.33. The minimum value is clearly 0.64150, corresponding to
the decision u=-1.

Table 3.33   Calculations at x=3, k=3

u	g(k,u,k)	I(g,k+1)	L(x,u,k)	Total Cost
1	3.5	X	X	X
0	3	1.54979	0.09958	1.64937
-1	2.5	0.59171	0.04979	0.64150

For x=x, k=2, the calculations are very similar. We
attempt to find an interpolation formula of the form

$$I(x,3) = a + bx + cx^2$$

by obtaining an exact fit at x=1, x=2, and x=3.

The equations to be solved are

$$a + b + c = 1.23898$$
$$a + 2b + 4c = 0.40601$$
$$a + 3b + 9c = 0.64150$$

The solution is

$$a = 3.14041$$
$$b = 2.43566$$
$$c = 0.53423$$

Substituting these values for x=2.5, we obtain

$$I(2.5,3) = 0.39020$$

The complete calculations for x=3, k=2 are shown in Table 3.34.

Table 3.34   Calculations at x=3, k=2

u	g(x,u,k)	I(g,k+1)	L(x,u,k)	Total Cost
1	3.5	X	X	X
0	3	0.64150	0.09958	0.74108
-1	2.5	0.39020	0.04979	0.43999

The minimum value is clearly 0.43999, corresponding to the decision
u=-1.

The procedure continues in this fashion until all results for
x=0,1,2,3 and k=0,1,2,3,4,5 have been obtained. As noted above,
the complete results are shown in Figure 3.20.

(b)   For the initial state x(0) = 2, the optimum solution is unchang
from the previous problem, since the optimum decision policy does
not change when the quadratic interpolation is used. The solution
is thus as shown in Table 3.32.

It is instructive to note that the cost along this trajectory,
0.96219, is much closer to the value calculated for this trajectory
using quadratic interpolations, 0.93960, than to the value calculated
using linear interpolation, 1.14363. The improvement in accuracy
is by nearly a factor of 6. Thus, in this highly nonlinear example
there is a major payoff for using the higher-order interpolation
formula.

3.6   Now, let us consider a problem with two state variables, x_1 and x_2, and two decision variables, u_1 and u_2. The performance criterion is

$$J = \sum_{k=0}^{3} |x_1(k)| + |x_2(k)| + 0.5|u_1(k)| + 0.5|u_2(k)|$$

$$+ |x_1(4)| + |x_2(4)|.$$

The system equations are

$$x_1(k+1) = x_1(k) + x_2(k) + u_1(k)$$

$$x_2(k+1) = -x_1(k) + x_2(k) + u_2(k)$$

The constraints are

$$-2 \leq x_1(k) \leq 2$$
$$-2 \leq x_2(k) \leq 2$$
$$-1 \leq u_1(k) \leq 1$$
$$-1 \leq u_2(k) \leq 1$$

(a) Apply the basic dynamic programming computational procedure to obtain a complete solution for the problem with quantization increments

$$\Delta x_1 = \Delta x_2 = \Delta u_1 = \Delta u_2 = 1.$$

(b) Determine the optimal trajectory and optimum decision sequence from $x_1(0) = 1$, $x_2(0) = 2$.

(c) Compare the computational requirements for this problem with those of a one-dimensional example with the same quantization increments in both the state variable and the decision variable.

(a)   We must now use a two-dimensional grid to represent the
state space at each stage. The allowed quantized states are
clearly

$$x = \{(-2, -2), (-2,-1), (-2,0), (-2,1), (-2,2), (-1,-2),$$
$$(-1,-1), (-1,0), (-1,1), (-1,2), (0,-2), (0,-1),$$
$$(0,0), (0,1), (0,2), (1,-2), (1,-1), (1,0), (1,1),$$
$$(1,2), (2,-2), (2,-1), (2,0), (2,1), (2,2)\}.$$

The terminal costs, $|x_1| + |x_2|$, are entered above and to the right
of the grid points in Figure 3.21.

The computational procedure now proceeds as in the usual
manner, except that we must keep track of two state variables and
two decision variables. The allowed quantized values of the decision variables are

$$U = \{(-1,-1), (-1,0), (-1,1), (0,-1), (0,0), (0,1),$$
$$(1,-1), (1,0), (1,1)\}.$$

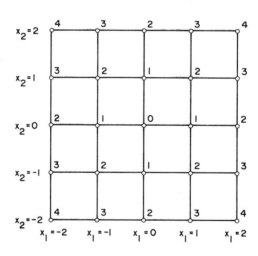

Figure 3.21   Terminal Costs in Solved Problem 3.6

The calculations at $x_1(3) = 0, x_2(3) = 2$ are summarized in
Table 3.35. The x's in the table indicate that the resulting
next state does not satisfy the constraints. The remaining
calculations at k=3 can be completed using this table. The results
are shown in Figure 3.22. The optimum value of u_1 is below and
to the left at the grid point, while the optimum value of u_2 is
below and to the right of the grid point. The x's at the grid
points (-2,-2), (-2,2), (2,-2), and (2,2) indicate that there is
no quantized admissible decision for which the resulting next
state satisfies the constraints.

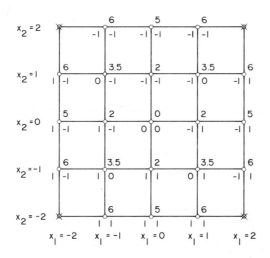

Figure 3.22   Solution for k=3 in Solved Problem 3.6

The calculations proceed in a similar manner for k=2, k=1, and
k=0. The results at these stages are shown in Figures 3.23, 3.24
and 3.25, respectively.

Table 3.35   Calculations at $x_1 = 0$, $x_2 = 2$, $k = 3$

u_1	u_2	$x_1 + x_2 + u_1$	$-x_1 + x_2 + u_2$	$I(g, k+1)$	$L(x_1, x_2, u_1, u_2, k)$	SUM
-1	-1	1	1	2	3	5
-1	0	1	2	3	2.5	5.5
-1	1	1	3	X	X	X
0	-1	2	1	3	2.5	5.5
0	0	2	2	4	2	6
0	1	2	3	X	X	X
1	-1	3	1	X	X	X
1	0	3	2	X	X	X
1	1	3	3	X	X	X

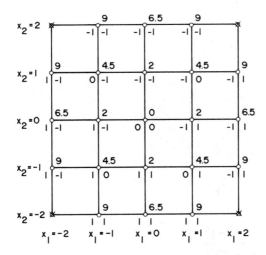

Figure 3.23   Solution for k=2 in Solved Problem 3.6

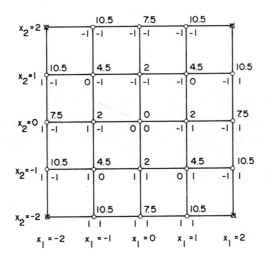

Figure 3.24   Solution for k=1 in Solved Problem 3.6

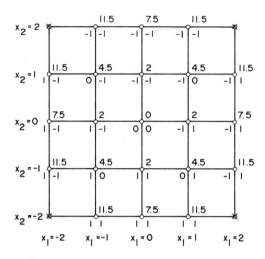

Figure 3.25   Solution for k=0 in Solved Problem 3.6

(b)   The optimal trajectory and optimum decision sequence can be recovered in a straightforward manner. First $\hat{u}_1(\hat{x}_1(k),\hat{x}_2(k),1)$ and $\hat{u}_2(\hat{x}_1(k),\hat{x}_2(k),k)$ are recovered from the grid. Next, $\hat{x}_1(k+1)$ and $\hat{x}_2(k+1)$ are determined from

$$\hat{x}_1(k+1) = \hat{x}_1(k) + \hat{x}_2(k) + \hat{u}_1(\hat{x}_1(k),\hat{x}_2(k),k)$$

$$\hat{x}_2(k+1) = -\hat{x}_1(k) + \hat{x}_2(k) + \hat{u}_2(\hat{x}_1(k),\hat{x}_2(k),k)$$

The calculations begin at $\hat{x}_1(0) = 1$, $\hat{x}_2(0) = 2$ and continue until $\hat{x}_1(4)$ and $\hat{x}_2(4)$ are found. The results are shown in Table 3.36.

Table 3.36

k	$\hat{x}_1$	$\hat{x}_2$	$\hat{u}_1$	$\hat{u}_2$	$L(\hat{x}_1,\hat{x}_2,\hat{u}_1,\hat{u}_2)$
0	1	2	-1	-1	4
1	2	0	-1	1	3
2	1	-1	0	1	2.5
3	0	-1	1	1	2
4	0	0	--	--	0
SUM	--	--	--	--	11.5

(c)   In the two-dimensional problem there are 25 quantized states
and 9 quantized decisions. In a one-dimensional example with the
same quantization increments there would be only 5 state variables
and 3 decision variables. Using the formulas in the text, the
high speed memory requirements is thus increased from 5 to 25; the
number of basic computations is increased from $5 \cdot 3 \cdot 5 = 75$ to
$25 \cdot 9 \cdot 5 = 1125$; and the number of results obtained increases from
$5 \cdot 5 \cdot 2 = 50$ to $25 \cdot 5 \cdot 3 = 375$. The increases are clearly very
substantial; if the numbers are divided by 5 to obtain a comparison
of the computational effort per stage, the increase is seen to
be basically a quadratic increase in effort.

3.7   Consider the resource allocation problem of maximizing the
multiplicative performance criterion

$$J = \prod_{k=1}^{5} (1+ku(k))$$

where u(k), the resource allocated at stage k, is constrained by

$$0 \leq u(k) \leq 5,$$

and where the total resource available is limited by

$$\sum_{k=1}^{5} u(k) = 8$$

Solve this problem numerically using quantization increment $\Delta u = 1$.

We first define the state variable, x(k), to be the amount
of resource left to allocate to stages k through 5 inclusive. The
system equation then becomes
$$x(k+1) = x(k) - u(k).$$
Recalling the basic iterative functional equation for a multiplicative criterion from Chapter 1, we see that the equation to be
solved is

$$I(x,k) = \underset{u}{Max}\{(1+ku)\ I(x-u,k+1)\}$$

where $I(x,k)$ is the optimal return from allocating x units of
resource to stages k, k+1,...,5. The quantized admissible values
of the decision variable become $U = \{0,1,2,3,4,5\}$, while the
quantized admissible values of the state are $X = \{0,1,2,3,4,5,6,7,8\}$.
The terminal condition is clearly

$$I(x,5) = 1+5x,\ 0 \leq x \leq 5$$
$$= 26,\ \ x = 6,7,8$$

The complete results are shown in Figure 3.26. The solution is
$u(1) = 1,\ u(2) = 1,\ u(3) = 2,\ u(4) = 2,\ u(5) = 2$, with a resulting
value for the performance criterion of 4,158.

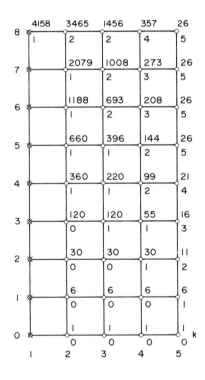

Figure 3.26   Complete Results for Solved Problem 3.7

3.8   As another specific example of a resource allocation problem, consider the problem faced by an engineer in planning how to spend his time over the day. He finds that he is responsible for four major tasks: working on existing assignments, planning new programs, preparing presentations to his superiors, and administration. Based on years of experience, he has determined his effectiveness for each task as a function of number of hours spent on the task. This effectiveness measure for each task is shown in Table 3.37. Experience has shown that he must spend at least one hour and no more than five hours on each task.

(a)   Find the best allocation of his eight-hour day to these four tasks and his corresponding total effectiveness.

(b)   If he is paid one hour of overtime, how should he spend the extra hour and what is his increased effectiveness?

Table 3.37   Effectiveness Measure for Each Task of Engineer

UNITS OF EFFECTIVENESS				
HOURS SPENT	WORKING	PLANNING	PRESENTATIONS	ADMINISTRATION
1	5	6	8	7
2	12	11	13	10
3	17	15	16	12
4	20	18	18	13
5	22	20	19	14

(a)   Using the results of the previous chapter, we see that the recursive equation to be solved is

$$I(x,k) = \min_{u}\{R(u,k) + I(x-u, k+1)\}$$

where I(x,k) = total effectiveness obtained by allocating x hours
               to tasks k, k+1,...,4, where k=1 is identified as
               working, k=2 as planning, k=3 as presentations and
               k=4 as administration;
      R(u,k) = effectiveness obtained by allocating u hours
               to task k; these values correspond to the entries
               in Table 3.37.

From the problem constraints, we see that u must take on one of the
values in U = {1,2,3,4,5}. Since we are interested in answers
for both 8 hour and 9 hour days, the set of admissible states is
taken as X = {1,2,3,4,5,6,7,8,9}. The equation is solved for
k=1,2,3. The starting condition is simply

$$I(x,4) = R(x,4)$$

where R(x,4) is the set of entries under task 4, administration.

     The complete dynamic programming solution is shown in Table
3.38. The state x, corresponding to total hours remaining to be
allocated, is tabulated vertically. The stage k, corresponding to
each task, is represented by the columns. For each state and stage,
x and k, the entry in this table contains the optimum effectiveness
I(x,k), followed by the optimum amount of resource to allocate on
this task, û(x,k). From this table, we see that the optimum
effectiveness of allocating 8 hours to all tasks is given by
I(8,1) = 48 units. The corresponding allocation to each task is
found by tracking back from this state and stage; the points along
this sequence are enclosed in rectangles. We see, then, that the
optimum allocation of 8 hours is 3 hours to working, 2 hours to
planning, 2 hours to presentations, and 1 hour to administration.

(b) From the table we see that the optimum effectiveness of
allocating 9 hours to all tasks is given by I(9,1) = 52 units. The
marginal return in effectiveness for the extra hour of overtime is
thus seen to be four units. The corresponding allocation to each
task is found by tracing back from state 9, stage 1; the points

Table 3.38   Complete Dynamic Programming Solution

STAGE, k STATE x	WORKING	PLANNING	PRESENTATIONS	ADMINISTRATION
9	(52,3)	X	X	X
8	48,3	41,4	X	X
7	43,3	38,4	30,4	X
6	38,3	(35,3)	28,4	X
5	33,2	31,2	26,3	14,5
4	26,1	26,2	23,3	13,4
3	X	21,1	(20,2)	12,3
2	X	X	15,1	10,2
1	X	X	X	(7,1)

Each entry in the table is $I(x,k)$, $\hat{u}(x,k)$.

along this sequence are enclosed in ovals. We see, then, that the optimum allocation of 9 hours is 3 hours to working, 3 hours to planning, 2 hours to presentations, and one hour to administration. Thus, the additional overtime hour is optimally utilized for planning.

3.9   Let us use the results of Solved Problem 2.6 to solve the unit commitment problem for the simplified power system shown there. As noted before, Unit 1 runs at all times. The cost of starting either Unit 2 or Unit 3 is $25, while the cost of shutting down either unit is 0. Each unit has a capacity of 10. The system day begins at time 0 with Unit 1 on and demand $D(0) = 0$, and it ends at time 5 with Unit 1 on and demand $D(5) = 0$. The demand at other times is given in Table 3.39. Use dynamic programming and the economic dispatching solutions, $f_1(D)$, $f_2(D)$, and $f_3(D)$, to find the start-up costs and operating costs. Also, specify the minimum total cost.

Table 3.39   System Demand

k	D(k)
0	0
1	5
2	10
3	15
4	10
5	0

Recall that $f_1(D) = \frac{1}{2}D^2 = 0.500D^2$, $f_2(D) = \frac{1}{3} D^2 = .333 D^2$ and $f_3(D) = \frac{3}{11} D^2 = .273 D^2$.

In the unit commitment problem, both start-up costs and operating costs are considered to determine which units are utilized at each hour of the day and how the demand is distributed over these units. These composite cost curves, $f_1(D)$, $f_2(D)$, and $f_3(D)$, tell us the operating costs for 1, 2, and 3 units on. Now all we need is a framework for deciding which units to start-up and shut-down at each stage.

A little thought tells us that the state variable for this problem is the number of units on, i.e.,

$$x(k) = 1, 2, \text{ or } 3$$

where $x(k) = j$ signifies that j units are on. If we know $x(k)$, we can compute the operating costs at each stage and determine start-up costs at each stage for any change in number of units on.

Clearly, the decision variable is the change in number of units on. Formally,

$$u(k) = -2, -1, 0, 1, \text{ or } 2$$

where $u(k) = 0$ means no start-up or shut-down, $u(k) = 1$ or 2 means 1 or 2 units are started-up, while $u(k) = -1$ or -2 means 1 or 2 units are shut-down.

The system equation is

$$x(k+1) = x(k) + u(k)$$
$$x(0) = 1$$

The admissible states and decisions are already quantized, viz.

$$X = \{1, 2, 3\}$$
$$U = \{-2, -1, 0, 1, 2\} \ .$$

The cost function combines start-up costs and operating costs. The operating cost can be written

$$L_1(x,k) = f_x(D(k)), \quad x = 1,2,3$$

The start-up costs can be summarized as

$$L_2(u,k) = 50, \quad u = 2$$
$$25, \quad u = 1$$
$$0, \quad u = 0, -1, -2.$$

The total cost at each stage is thus

$$L(x,u,k) = L_1(x,k) + L_2(u,k)$$

Because everything is already quantized, we can set up our usual grid and proceed directly. Since $x(5) = 1$, $D(5) = 0$, and there is no decision $u(5)$, we obtain the starting condition.

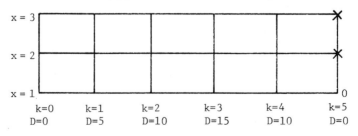

Figure 3.27   Initial Grid for Solved Problem 3.9

The calculations can be carried out relatively rapidly by pre-computing the functions $L_1(x,k)$ for the allowed values of demands, $D = 0, 5, 10$ and 15.

$$f_1(0) = f_2(0) = f_3(0) = 0$$

$$f_3(5) = 6.8$$
$$f_2(5) = 8.3$$
$$f_1(5) = 12.5$$

$$f_3(10) = 27.3$$
$$f_2(10) = 33.3$$
$$f_1(10) = 50.0$$

$$f_3(15) = 61.4$$
$$f_2(15) = 75.0$$
$$f_1(15) = 111.5$$

The final grid is as shown in Figure 3.28.

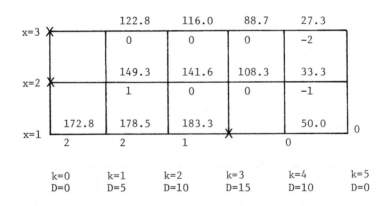

Figure 3.28   Final Grid for Solved Problem 3.9

The number above and to the right of the grid point is I(x,k) and
the number below and to the right is û(x,k). The x at x=1, k=3
occurs because the demand 15 exceeds the capacity of a single unit.
The x's at x=2, k=0 and x=3, k=0 occur because the initial state is
fixed at x(0) = 1.

We can see directly that the minimum cost is 172.8. The optimal
decision policy is to turn on all three units at time 1 and to shut
them down at time 5. The solution is summarized in Table 3.37.

Table 3.40   Solution to Solved Problem 3.9

k	x	u	L_1	L_2	L
0	1	2	0	50	50
1	3	0	6.8	0	6.8
2	3	0	27.3	0	27.3
3	3	0	61.4	0	61.4
4	3	-2	27.3	0	27.3
5	1	--	0	0	0
			172.8	50	172.8

3.10 Consider the following electric utility expansion planning
problem. New units are available in two sizes, 500 MW and 1,000 MW.
One new unit of either type can be bought in 1980, 1985, and 1990.
There is an economy-of-scale advantage to the larger unit, and a
significant discount rate is assumed. The purchase costs, normalized
to 1975 dollars, are shown as a function of time and unit size in
Table 3.41.

Table 3.41   Purchase Cost, 1975 dollars $(\times 10^{+6})$

Unit Size \ Year	1980	1985	1990
500 MW	1.0	0.5	0.25
1000 MW	1.6	0.8	0.4

The system load is increasing over the planning period. The cumulative increase in system peak load is as shown in Table 3.42.

Table 3.42   Cumulative Increase System Peak
             Load Over 1975

YEAR	CUMULATIVE INCREASE
1980	400 MW
1985	800 MW
1990	1200 MW

If the cumulative capacity of new units installed exceeds cumulative increase in peak demand, operating cost is assumed to be normal. If cumulative capacity falls short by 500 MW or more, the system is not feasible. If the cumulative capacity falls short by an amount less than 500 MW, then it is assumed that the system can still operate, but with a penalty in operating cost given in Table 3.43.

Table 3.43   Total Operating Cost Penalty for
             Cumulative New Capacity Falling Short
             of Cumulative Increase in Demand,
             1975 Dollars

Year of Shortfall / Shortfall	1980	1985	1990
100	0.2	0.1	0.05
200	0.4	0.2	0.1
300	0.6	0.3	0.15
400	0.8	0.4	0.2

Find the optimal expansion plan for this system, i.e., determine what size unit (if any) to build in each of years 1980, 1985 and 1990, such that the total capacity cost and total operating cost penalty is minimized. Assume that at most one unit can be purchased in any one year.

This problem has a structure similar to that of a resource allocation problem. The stage variable for this problem can be taken as k=0 for 1975, k=1 for 1980, k=2 for 1985, and k=3 for 1990. The state variable, $x(k)$, is the total new capacity installed by stage k. The decision variable, $u(k)$, is the amount of new capacity installed just after stage k. The system equation thus becomes

$$x(k+1) = x(k) + u(k), \quad k = 0,1,2$$

where the new capacity installed just after stage k is assumed to be available at stage k+1. The initial capacity is $x(0) = 0$.

The total capacity can be quantized in increments of 500 MW, because the capacity additions are either 500 MW or 1,000 MW. The maximum capacity that needs to be considered is 1,500 MW, which is the smallest quantized value that exceeds the 1990 demand of 1,200 MW. The set of admissible states is thus $X = \{0, 500, 1000, 1500\}$. The set of admissible decisions is $U = \{0, 500, 1000\}$.

The cost function includes two terms, one for capacity additions and one for the penalty due to total capacity falling short of demand. The capacity addition cost can be calculated from Table 3.41, where the capacity installed just after stage k is assumed to be paid for at stage k+1. Note that this term depends only on the decision variable. If this cost is denoted as $L_1(u,k)$, it can be tabulated as in Table 3.44. The shortfall penalty term can be calculated using the peak load from Table 3.42 and the penalty data in Table 3.43. Note that this term depends only on the state variable. If this cost is denoted as $L_2(x,k)$, it can be tabulated as in Table 3.45.

The basic recursive equation thus can be written as

$$I(x,k) = \min_u \{L_1(u,k) + L_2(x,k) + I(x+u, k+1)\}.$$

$$k = 0,1,2$$

$$I(x,3) = L_2(x,3).$$

192

LARSON AND CASTI

Table 3.44   Capacity Addition Cost, $L_1(u,k)$

k u	0	1	2
0	0.0	0.0	0.0
500	1.0	0.5	0.25
1000	1.6	0.8	0.4

Table 3.45   Shortfall Penalty Costs, $L_2(x,k)$

k x	0	1	2	3
0	0.0	0.8	X	X
500	0.0	0.0	0.3	X
1000	0.0	0.0	0.0	0.1
1500	0.0	0.0	0.0	0.0

X denotes inadmissible state.

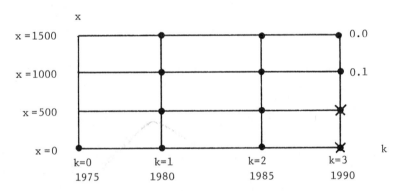

Figure 3.29   Initial Grid

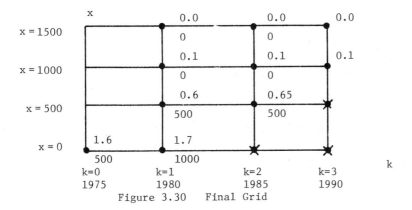

Figure 3.30   Final Grid

The solution is summarized in Table 3.46.

Table 3.46   Solution to Problem 3.9

k	$\hat{x}(k)$	$\hat{u}(k)$	$L_1(\hat{u},k)$	$L_2(\hat{x}+\hat{u},k+1)$
0	0	500	1.0	0.0
1	500	500	0.5	0.0
2	1000	0	0.0	0.1
3	1000	--	--	--

Total Cost = 1.6

SUPPLEMENTARY PROBLEMS

3.11   Consider the problem with performance criterion

$$J = \sum_{k=0}^{4} [x^2(k) + u^2(k)] + 2.5(x(5) - 2)^2,$$

system equation

$$x(k+1) = x(k) + u(k),$$

constraints

$$0 \leq x \leq 2 ,$$
$$-1 \leq u \leq 1 ,$$

and initial condition

$$x(0) = 2.$$

The complete dynamic programming solution to this problem with quantization increments $\Delta x = 1$ and $\Delta u = 1$ was obtained in Problem 3.1.

(a)   Solve this problem for the specified initial state using the brute-force enumeration procedure.   Take as the set of admissible decisions the same set as for the dynamic programming procedure, $U = \{-1,0,1\}$.   The complete decision tree for this example was traced out in Figure 3.2;   all that is required here is to fill in the value of $\Omega(x(k),k)$ for all $x(k)$ indicated on this tree.

(b)   Find the solution(s) for this problem by examining the completed decision tree.   Would you expect the solution to be the same as obtained for this initial state by dynamic programming in Example 3.1?   Verify your answer directly and explain the result.

(c)   Use the enumeration procedure to find the solution for initial state $x(2) = 1$.   Can the solution for $x(0) = 2$ be used at all?

Explain and compare the solution for x(2) = 1 with that obtained by dynamic programming in Example 3.1.

(d)   Use the enumeration procedure to find the solution for an initial state x(0) = 0. Can the solution for x(0) = 2 be used at all? Explain. Compare the solution for x(0) = 0 with that obtained by dynamic programming in Example 3.1.

3.12   Consider the problem with performance criterion

$$ J = \sum_{k=0}^{9} [5x^2(k) + u^2(k)] + 2.5[x(10) - 2]^2, $$

system equation

$$ x(k+1) = x(k) + u(k), $$

and constraints

$$ 0 \le x(k) \le 8, \quad k = 0,1,\ldots,9 $$
$$ 0 \le x(10) \le 2 $$
$$ -2 \le x(k) \le 2, \quad k = 0,1,\ldots,9 $$

Use uniform quantization increments $\Delta x=1$ and $\Delta u=1$. Note the similarity of this problem to the illustrative example worked out in this chapter and to Problem 3.3.

(a)   Find the complete dynamic programming solution to this problem.

(b)   Trace out the solution from x(0) = 8. Compare the solution to that for x(0) = 8 in the illustrative example and in Problem 3.3. Explain any differences.

(c)   Trace out the solution from x(3) = 4. Compare the solution to that for x(3) = 4 in the illustrative example and in Problem 3.3. Explain any differences.

3.13   Consider the problem with performance criterion

$$J = \sum_{k=0}^{4} (2 + u(k)) \, e^{-x(k)} + |x(5) - 1| \, ,$$

system equation

$$x(k+1) = x(k) + [2 - 2x(k) + \frac{5}{4} x^2(k) - \frac{1}{4} x^3(k)] \, u(k) \, ,$$

and constraints

$$0 \leq x \leq 3 \, ,$$
$$-1 \leq u \leq 1 \, .$$

A complete dynamic programming solution for this problem with quantization increments $\Delta x = 1$ and $\Delta u = 1$ was found in Problem 3.4 using linear interpolation for both $I(g, k+1)$ and $\hat{u}(x, k)$, and in Problem 3.5 using quadratic interpolation $I(g, k+1)$ and linear interpolation in $\hat{u}(x, k)$. Now, obtain the complete dynamic programming solution and trace out the solution from initial state $x(0) = 2$ for quantization increments $\Delta x = \frac{1}{2}$, $\Delta u = \frac{1}{2}$ using linear interpolation for all interpolations required. Compare these results with those obtained in Problems 3.4 and 3.5. Explain the trade-off between using finer quantization increments and more accurate interpolation formulas in terms of these results.

3.14   Consider the problem with performance criterion

$$J = \sum_{k=0}^{10} [x^2(k) + u^2(k)] \, ,$$

system equation

$$x(k+1) = x(k) - u(k) \, ,$$

constraints

$$0 \leq x \leq 4 \, ,$$
$$0 \leq u \leq 1 \, ,$$

and initial condition

$\quad$ x(0) = 4.

Quantize in such a way that

$\quad$ X = {0,1,2,3,4}

$\quad$ U = {0,0.1,0.2,0.3,0.4,0.5,0.6,0.7,0.8,0.9,1.0}

Develop the complete dynamic programming solution and recover the optimal trajectory from x(0) = 4 under the following interpolation schemes:

(a)$\quad$ Use the value for the nearest quantized state x for both $I(x,k)$ and $\hat{u}(x,k)$.

(b)$\quad$ Use linear interpolation between values at the two nearest quantized states x for both $I(x,k)$ and $\hat{u}(x,k)$.

(c)$\quad$ Use linear interpolation between values at the two nearest quantized states x for $\hat{u}(x,k)$ and use quadratic interpolation between values at three nearest quantized states x for $I(x,k)$.

(d)$\quad$ Comment on the relative accuracy and computational requirements of these interpolation schemes.

3.15$\quad$ Consider the problem with performance criterion

$$J = \sum_{k=0}^{4} [x_1^2(k) + x_2^2(k) + u_1^2(k) + u_2^2(k)]$$

$$+ 2.5 [x_1(5) - 2]^2 + 2.5 [x_2(5) - 1]^2 ,$$

system equations

$\quad$ $x_1(k+1) = x_1(k) + x_2(k) + u_1(k)$,

$\quad$ $x_2(k+1) = x_2(k) + u_2(k)$,

and constraints

$$0 \leq x_1 \leq 2,$$

$$-1 \leq x_2 \leq 1,$$

$$-1 \leq u_1 \leq 1,$$

$$-1 \leq u_2 \leq 1.$$

Quantize the state variables in uniform increments of $\Delta x_1 = \frac{1}{2}$ and $\Delta x_2 = \frac{1}{4}$ respectively. Do not quantize the control variables, but allow them to take on any admissible value such that for any given quantized present state the next state is also a quantized state. This eliminates the need for interpolation.

(a) Find the complete dynamic programming solution to this problem.

(b) Trace out the optimal solution from the initial state $x_1(0)=2$, $x_2(0) = 1$.

(c) Trace out the optimal solution from the initial state $x_1(0)=0$, $x_2(0) = -1$.

3.16 Consider the problem with performance criterion

$$J = \sum_{k=0}^{4} [x_1^2(k) + x_2^2(k) + u_1^2(k) + u_2^2(k)],$$

system equations

$$x_1(k+1) = 0.625x_1(k) + 0.25x_2(k) + u_1(k),$$

$$x_2(k+1) = -0.1875x_1(k) + 0.125x_2(k) + u_2(k),$$

and constraints

$$-5 \leq x_1 \leq 5 ,$$

$$-5 \leq x_2 \leq 5 ,$$

$$-2 \leq u_1 \leq 2 ,$$

$$-2 \leq u_2 \leq 2 .$$

Quantize the state variables in uniform increments of $\Delta x_1 = \frac{1}{2}$ and $\Delta x_2 = \frac{1}{2}$ respectively. As in the previous example, do not quantize the control variables, but instead determine them such that for any given quantized present state the next is also a quantized state.

(a)   Find the complete dynamic programming solution to this problem.

(b)   Trace out the optimal solution from the initial state $x_1(0)=5$, $x_2(0) = 5$.

(c)   Trace out the optimal solution from the initial state $x_1(0)=5$, $x_2(0) = -5$.

3.17   Consider the problem of minimum fuel trajectories for a jet aircraft in its cruise region. The forces on the aircraft when it is flying under steady-state conditions at a flight path angle can be deduced from Figure 3.31. The forces on the aircraft can be resolved into components along and perpendicular to the direction of flight as follows:

$$F_v - \frac{W}{g} \frac{dv}{dt} = T - D - W\sin \gamma$$

$$F_p = 0 = L - W\cos \gamma$$

where

   F_v = Force along flight path

   F_p = Force perpendicular to flight path

   W = Weight of aircraft, including fuel

   g = Acceleration due to gravity

   v = Velocity of aircraft

   h = Altitude

   T = Engine thrust

   D = Drag

   L = Lift

   γ = Flight path angle with respect to horizontal

   r = Range

These dynamics for the aircraft can be summarized in the differential
equations.

$$\frac{dv}{dt} = \frac{g}{W} \; (T - D - W \sin \gamma)$$

$$\frac{dh}{dt} = v \sin \gamma$$

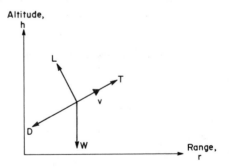

Figure 3.31   Force Diagram on Aircraft

The fact that the aircraft consumes fuel as it flies can be
expressed as

$$\frac{dW}{dt} = -W_F$$

where W_F is the fuel flow for the engines. The fuel flow is a
function of thrust, weight, velocity, and altitude.

$$W_F = f_1(T,W,v,h).$$

Similarly, drag can be written as a function of lift, weight,
velocity and altitude.

$$D = f_2(L,W,v,h) \ .$$

Finally, there are constraints limiting the values of thrust,
flight path angle, velocity, and altitude.

$$T^- \leq T \leq T^+$$

$$\gamma^- \leq \gamma \leq \gamma^+$$

$$v^- \leq v \leq v^+$$

$$h^- \leq h \leq h^+$$

(a)   Assume that the aircraft flies at known constant velocity,
v^*. Assume further that the weight of the aircraft, including
fuel, is known at the <u>end</u> of the flight as W_T. Then, identifying
the state variable x as altitude h, the control variable u
as flight path angle γ, and the stage variable t as range r,
show that the problem of determining the minimum fuel trajectory
problem is equivalent to minimizing the performance criterion

$$J = \int_{r_o}^{r_f} f_1[f_2(W \cos u, W, v^*, x) + W \sin u, W, v^*, x] \, dt$$

subject to the scalar system equation

$$\frac{dx}{dt} = \tan u$$

the constraints

$$\gamma^- \le u \le \gamma^+$$

$$h^- \le x \le h^+$$

$$T^- \le f_2 \ (W \cos u, \ W, \ V^*, \ x) + W \sin u \le T^+,$$

the initial condition

$$x(r_o) = h_o \ ,$$

the terminal constraints

$$x(r_f) = h_f \ ,$$

and the side condition

$$W = W_T + J \ ,$$

where r_o = initial range of aircraft

r_f = final range of aircraft

h_o = initial altitude of aircraft

h_f = final altitude of aircraft

(b)   Assume that for normalized aircraft parameters the performance criterion becomes

$$J = \sum_{k=0}^{19} \ 0.05 \ W(x,u,k) \ [3e^{-0.15x(k)} \ \{1 + 0.5 \ \sin^2 u(k)\}$$

$$+ \sin u(k)] \ ,$$

and the system equation takes the form

$$x(k+1) = x(k) + \tan u(k) \cdot$$

The quantity $W(x,u,k)$, which is the weight of the aircraft at state x, stage k when control u is applied, is taken to be

$$W(x,u,k) = W_T + I(x + \tan u, \ k+1)$$

where $I(x + \tan u, k+1)$ = minimum fuel required to take aircraft
to end from state $x + \tan u$, stage $k+1$.

The terminal weight of the aircraft is assumed to be

$$W_T = 1 \,,$$

the initial state is taken as

$$x(0) = 5,$$

and the terminal state is

$$x(20) = 0.$$

The constraints are

$$0 \leq x \leq 10$$
$$-45° \leq u \leq 45°$$

Solve this problem using dynamic programming. Quantize the state
and control variables in uniform increments $\Delta x = 1$ and $\Delta u = 15°$
respectively. If interpolation is required, use a linear interpolation between the two nearest quantized values.

3.18 Consider the simplified hydro-thermal power system shown in
Figure 3.32. The total power demand must be satisfied by one
hydro generating station and several thermal generating stations.
Because the demand varies over the day and because the thermal
stations have nonlinear cost characteristics, the value of hydro
generation varies during the day. During times of low demand, only
efficient thermal plants need be utilized, and the cost savings by
using hydro instead of thermal generation are small. On the other
hand, during periods of high demand less efficient thermal plants
may be used, and the cost savings by using hydro instead of thermal
generation can be great. For this example, let $x(k)$ represent
the amount of water stored in the reservoir at time k, $u(k)$ the

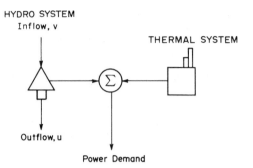

Figure 3.32   Simplified Hydro-Thermal Power System

amount of water released from the reservoir at time k, and v(k)
the amount of water flowing into the reservoir at time k, where
k=0,1,...,23 represents an index of hourly intervals over the day.
A mass balance on water in the reservoir shows that

$$x(k+1) = x(k) - u(k) + v(k) \; ,$$
$$k=0,1,...,23$$

The reservoir level is constrained to be within the interval

$$0 \leq x \leq 10.$$

The initial and final reservoir levels are taken to be

$$x(0) = x(24) = 5 \; ,$$

with the release limited by

$$0 \leq u \leq 3 \; \cdot$$

The inflow is constant at

$$v(k) = 2 \; ,$$
$$k=0,1,...,23 \; \cdot$$

Based on the nature of the power demand, the generation characteristics of the hydro stations and the cost characteristics of the
thermal system, the value of a release u(k) can be calculated as
the time-varying quantity c(k)u(k), where c(k) is given in Table
3.47.   The total value of the releases is thus

Table 3.47 $c(k)$, Value of One Unit of Hydro Release at
Hourly Interval k, $k=0,1,\ldots,23$

k	$c(k)$
0	1.2
1	1.1
2	1.0
3	1.0
4	1.0
5	1.1
6	1.2
7	1.4
8	1.7
9	2.0
10	2.3
11	2.5
12	2.4
13	2.5
14	2.6
15	2.5
16	2.4
17	2.2
18	2.6
19	2.3
20	2.0
21	1.7
22	1.5
23	1.3

$$J = \sum_{k=0}^{23} c(k)\ u(k)$$

If the releases are made in unit increments, determine the sequence
of releases over the day such that the total value of these releases
over the day is maximized and all other constraints are met.

3.19 An electrical engineering student at a major western university is about to take an important examination. The exam covers
questions from 10 categories, and he is given a grade of from 1 to
10 in each category. His expected score in each category is as
follows:

Control Theory	9
Computer Science	5
Communication Theory	7
Circuit Theory	5
Electronic Devices	4
Power Systems	4
Electromagnetic Theory	3
Physics	3
Mathematics	8
Fundamentals of Engineering	2

He is allowed to weight the score in each category by an integer
from 1 to 5 inclusive. The sum of the weights must equal 30.
For example, he may assign a weight of 3 to each of the ten categories, in which case his expected score is 150. Develop a dynamic
programming procedure to find the set of weights that maximizes
his expected score. Apply the procedure to find the optimum weights
for this case.

3.20   Consider the expansion planning problem for the communication network shown in Figure 3.33.   The capacities of the links in 1977 are as shown in Figure 3.33.   One unit of capacity can be added to each link; the cost of adding one unit of capacity to links 1-2 and 2-3 is $1.0M in 1978, while the cost of adding one unit of capacity to link 1-3 is $1.2M in 1978.   The discount rate is such that the cost of each addition in 1978 dollars decreases by $0.1M each year.

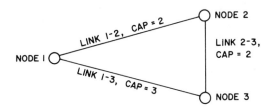

Figure 3.33   System Configuration in 1977

The discounted capital costs are thus as shown in Table 3.48.

Table 3.48   Capital Costs in 1977 Dollars for
Adding One Unit to Each Link

Link \ Year	1978	1979	1980
1-2	1.0	0.9	0.8
1-3	1.2	1.1	1.0
2-3	1.0	0.9	0.8

The traffic demands between nodes are as shown in Table 3.49.

Table 3.49   Traffic Demands Between Nodes

Node Pairs \ Year	1977	1978	1979	1980
Node 1 - Node 2	1.8	2.1	2.4	2.8
Node 1 - Node 3	2.2	2.5	2.8	3.2
Node 2 - Node 3	1.5	1.8	2.1	2.4

The total traffic demands can be met by a combination of direct link connections and communication via an intermediate node. However, the total traffic demand must be met and the total traffic on each link must be less than the link capacity. Also, the penalty for communicating via an intermediate node in any year is $0.1M in 1978 dollars. Find the optimal expansion plan for this system, i.e., specify the additions (if any) to each link in 1978, 1979, and 1980, such that total discounted capital costs and penalties for communications via an intermediate node are minimized, subject to the constraint that all traffic demands are met and all link capacity constraints are met. HINT: One potential state description is to use three states. with values 0 and 1, corresponding the presence or absence of the additional capacity at each link.

REFERENCES

Introduction: [B-41], [D-13], [J-1]

The Iterative Functional Equation
  vs. Direct Enumeration: [D-14], [N-4]

Constraints and Quantization: [F-21], [L-10], [L-15], [C-18]

Calculation of Optimal Control: [L-14]

Interpolation Procedures: [A-25], [D-9], [L-15], [S-13]

Computational Requirements: [B-6], [B-35], [W-24], [S-21],
                             [L-10], [L-8]

Chapter 4

EXTENSIONS OF THE BASIC PROCEDURES

INTRODUCTION

In the first three chapters of this volume, we have stressed the versatility and generality of dynamic programming. However, we have thus far focused on problems in which i) there is an explicit stage variable and ii) the basic recursive equations are solved backwards in this stage variable. In this chapter, we shall show that dynamic programming can be applied to problems in which one or both of these characteristics is absent. Furthermore, we will see that in many problems of practical importance there are significant computational advantages to adopting such a problem formulation, rather than always attempting to create an explicit stage variable and solving recursive equations backwards in this variable.

In the next few sections we shall discuss problems in which there is no explicit stage variable. Then we will examine problems in which there is an explicit stage variable, but an infinite number of stages; we shall develop procedures for solving such problems with a finite amount of computation. Finally, we will look at problems with a classical stage variable structure and a finite number of stages, but where the recursive equations of dynamic programming are solved forward in the stage variable. In all cases we will note the conceptual and computational advantages of these new viewpoints.

PROBLEMS WITH AN IMPLICIT STAGE VARIABLE

As we saw in Chapter 2, the critical step in dynamic programming is to find an imbedding of the original problem within a family
of problems such that
(a) One member of the family of problems can be solved in a straightforward fashion.
(b) Simple relations may be obtained linking various members of the
family of problems.

We saw that the stage variable was a convenient mechanism for
indexing the relations between problems. In this section we will
adopt a more general viewpoint allowing us to dispense with the
stage variable as an explicit entity.

Recall that in Chapter 1 we introduced the concept of a state
transition function in a multistage decision process. Such a function
is, in its most abstract sense, simply a mapping from one set of
states onto another set of states that depends on the values of a
set of decision variables. Formally, if X_D is the domain of the
transformation, i.e., if X_D is the set of states upon which the mapping is defined, if X_R is the set of states that result from the
transformation, and if U is the set of decisions that effect the
transformation, then the transformation g can be defined by specifying
for every $x \varepsilon X_D$ and every $u \varepsilon U$ that value of $x \varepsilon X_R$ that results from
the transformation. Formally if $c \varepsilon X_D$ and $u \varepsilon U$, then

$$g(c,u) = y, \qquad\qquad\qquad (4.1)$$

where $y \varepsilon X_R$.

In Chapters 1 and 2, we focused our attention on transformations that could be defined in terms of a stage variable. Let us
here take a more general viewpoint. However, we will still have to
place some restrictions on the transformation in order to achieve

results of significance. Let us now consider processes in which there is a subset of X_D upon which no further transformations can be made and the process can be considered to have ended. This subset is called the set of <u>terminal states</u> and denoted as X_T. Let us also assume that our process always begins from a subset of X_R called the set of <u>initial states</u> and denoted as X_I. Now we make the following definitions:

A <u>finitely terminable</u> process is defined to be a process in which

(a) There exists a transformation g and sets X_R, U and X_D such that for any $c \epsilon X_R$ and any $u \epsilon U$, a unique $y \epsilon X_D$ is defined.

(b) There exists a set X_T upon which no further transformations can be made and the process terminates.

(c) There exists a set X_I from which the process must always begin.

(d) For every state in the initial set X_I, every sequence of decisions from the set U will transform the state into a member of the terminal set X_T in a finite number of steps.

This concept is illustrated in Figure 4.1. If all possible decisions $u \epsilon U$ are applied for states $x \epsilon X_I$, then the resulting set of next states are $x \epsilon X(1)$. Note that some of these states, namely $X(1) \cap X_T$, are terminal states for which the process is finished and no further transformations are possible. Next, if all possible decisions $u \epsilon U$ are applied for nonterminal states in $X(1)$ (i.e., for $x \epsilon X(1) \cap \overline{X_T}$), then the resulting set of next states is $x(2)$. Again, some of these states are also terminal states, namely $x \epsilon X(2) \cap X_T$. Finally, if all possible decisions $u \epsilon U$ are applied for nonterminal states in $X(2)$, then the resulting set of next states is $X(3)$, which is completely contained in X_T. Thus, for this example all initial states will be mapped into terminal states in at most three transformations and the process is finitely terminable.

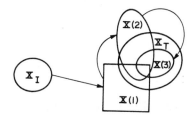

Figure 4.1   Illustration of a Terminable Process

Note that processes with an explicit stage variable and a finite number of states and stages are finitely terminable. In fact, a stage variable can be constructed for any finitely terminable process by indexing the number of transformations forward from the initial set of X_I until the point is reached where all states are in X_T; the number of transformations required to reach this condition defines the number of stages in the process. In this case it is also necessary to define a null transformation that applies no decision to any state in X_T that is reached prior to the last stage. As we shall see, however, there are some computational and conceptual advantages to solving the problem directly rather than reverting to this stage variable description.

Before applying dynamic programming to finitely terminable processes, one further restriction is required: the performance criterion must be Markovian in the sense of Chapter 1. The formal definition for these processes is a straightforward extension of that in Chapter 1 and will not be repeated here.

If the performance index is additively separable, i.e., if for any state x and decision u a transition cost is defined as $L(x,u)$, and if the total cost is the sum of these costs, then it is straightforward to write the basic dynamic programming equation as

$$I(x) = \underset{u}{\text{Min}} \ \{L(x,u) + I(g(x,u))\}$$ (4.2)

where I(x) is the minimum cost of completing the process from state
x. If the states $x \varepsilon X_T$ have a terminal cost $\phi(x)$ associated with
them, the calculations begin with

$$I(x) = \phi(x), \quad x \varepsilon X_T \qquad\qquad (4.3)$$

To solve these equations, a two-step procedure is required.
First, sets X(1), X(2),...,X(N) are defined by applying the transformation g for all decisions $u \varepsilon U$ states in X_I and continuing until
all states are in the terminal set X_T. Eq. (4.2) can now be solved
for the successive sets X(N-1), X(N-2),...,X(1), X_I, since at each
step the minimum cost will be known for all resulting next states.

The same basic procedure can be applied for any Markovian
performance criteria. This procedure eliminates the need to construct
an explicit stage variable. It also allows us to work with a minimal
set of states at each implicit stage. In this manner, the number of
states and stages is kept to the lowest possible value and significant computational advantages are obtained over a more brute-force
approach.

Unfortunately, there is no systematic procedure for determining
whether a particular problem can be treated as a finitely terminable
process rather than by defining an explicit stage variable, when
termination of the process is dependent on some particular condition
of these systems, not on the number of transformations that have
been applied.

The next few worked examples illustrate some cases where the
finitely terminable formulation is advantageous. Study of these
problems will help the reader develop skills in determining when
such a formulation should be attempted.

EXAMPLE

4.1 This example shows that dynamic programming can be applied to
problems where not only is there no explicit stage variable but the
system equations are not expressible as a function in closed form.

Let us consider the following water distribution network construction problem. Based on an analysis of water pressure zones and
topological considerations, a set of possible routings of pipe is
portrayed in Figure 4.2.   It is desired to find the least costly
routing of pipe from point A to point M.

The candidate pipe segments are represented in network theory
terminology as directed links between nodes, where the actual
physical locations of the nodes A,B,...,M are specified by the
designer. Each link has an arrow showing the direction water will
flow in the pipe.   Each link also has an associated total cost,
which includes all costs for materials (pipes, valves, etc.) and
construction (acquiring right of way, digging holes, connecting
pipes, etc.).   It is assumed here that sufficient analysis has been
done to select pipe parameters (diameter, wall strength, etc.)

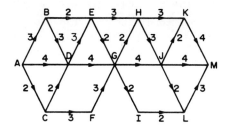

Figure 4.2   Possible Pipe Routings

such that all operating constraints are met for any connection of
pipes. The problem, then, is to find that connection of pipes
from A to M that has minimum total cost. Use dynamic programming
to formulate and solve this problem.

The system equations can be thought of here as a "state
transition function," which relates to the connection of one node
to another node by a link. The quantized "state" is thus the
index of the originating node. The quantized "decision" is the
index of the link that is installed. The "stage" is merely the
ordering of these connections. Thus, the "state-transition function"
expresses the next node along a given pipe routing as a function
of the present node and the link that is selected.

The single-stage cost function is simply the cost of the
link that is selected. The total cost function is the sum of
these single-stage costs.

The constraints have already been applied to specify the
"states" (nodes) that are admissible and the "decisions" (links)
that can be applied at any "state" (node).

To solve the recursive equation for the given network, we
note that the terminal condition for this case is

$$I(M) = 0$$

We next try to find nodes for which we can solve the recursive
equation. We see that nodes J, K and L all connect to node M.
We first try to find $I(J)$, the minimum cost of going to the terminal node from node J. This quantity is explicitly

$$I(J) = \min \ \{2 + I(K), \ 4 + I(M), \ 2 + I(L)\}$$

where the first term in each sum is the cost of the next link and
the second term is the resulting cost at the terminal end of the
link. Since only the value I(M) = 0 has been computed so far, we
cannot complete this calculation.

We next try to find I(K). This quantity is

$$I(K) = \min \; \{4 + I(M)\} \; = 4$$

Since there is only one link emanating from K, we find its
minimum cost immediately. We can portray this partial solution
in Figure 4.3 by putting the cost for the node in a circle and by
indicating with an arrow which link was selected.

Similarly, we find for node L that

$$I(L) = \min \; \{3 + I(M)\} = 3$$

This result is also shown in Figure 4.3.

We next proceed to consider all nodes that connect to any
link for which we have computed the minimum cost function· In
this case, the nodes for which we have values are K, L, and M, and
the nodes connected to them are H, I, and J.

At node H we see that

$$I(H) = \min \; \{3 + I(K), \; 3 + I(J)\}$$

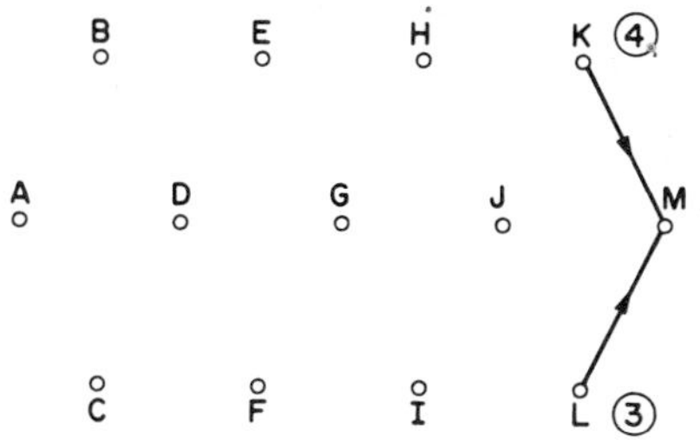

Figure 4.3   Results of Calculations After One Stage

This analysis cannot be completed because I(J) is not yet known. At node I we find

$$I(I) = \min \{2 + I(L)\}$$
$$= \min \{2 + 3\}$$
$$= 5.$$

The result is entered on Figure 4.4.

Proceeding to node J, we see that

$$I(J) = \min \{2 + I(K), \ 4 + I(M), \ 2 + I(L)\}$$
$$= \min \{2 + 4, \ 4 + 0, \ 2 + 3\}$$
$$= \min \{6, \ 4, \ 5\}$$
$$= 4$$

This result is also indicated on Figure 4.4.

The results of completing this procedure for the example are shown in Figure 4.5. From this solution we can trace back the optimum routing of the pipes as beginning at node A and passing through nodes B, E, G, I, and L before reaching the terminal node M. The cost of this path is 14 units.

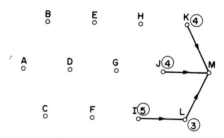

Figure 4.4   Results of Calculations After Two
             Stages

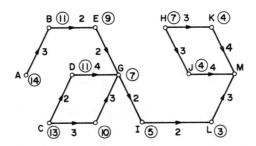

Figure 4.5.   Complete Solution for Example

As would be expected with a dynamic programming procedure, a "feedback" solution is obtained. The significance of this aspect of the results in this problem is that we have obtained the optimum routing from any initial node. For example, referring to Figure 4.5 there are two optimum routings from node C, one that passes through nodes D, G, I, L, and M and one through nodes F, G, I, L, and M. The cost of each routing is 13 units.

As an aside, it is worthwhile to note that in the context of this problem it is useful to retain both solutions and "break the tie" on the basis of non-economic factors, such as damage to the environment, inconvenience to citizens during construction, or projected reliability of the system.

EXAMPLE

4.2   Consider the problem of using an equal-arm balance to find the one heavy coin in a batch of N coins, the other N-1 of which have the same weight. Derive a dynamic programming recursive equation for the maximum number of weighings necessary to guarantee finding of the heavy coin.

Let us define the state variable x for this problem as the number of coins. Then,

I(x) = maximum number of weighings required to find
the heavy coin in a batch of x using an optimal
policy

The decision variable u in this problem corresponds to the number of coins placed on each side of the balance. If we weigh one batch of u coins against another batch of u coins, then clearly two outcomes are possible: (i) the two batches weigh the same, in which case the heavy coin is one of the remaining x−2u coins or (ii) the heavy coin is in one of the two batches of u coins. As a result of our choice to weigh u coins, we use up one weighing and apply an optimal policy from either the state u or the state x−2u. We thus see that

$$I(x) = 1 + \min_{1 \le u \le x/2} \{\max [I(u), I(x-2u)]\}, \quad I(1) = 0.$$

Clearly, to minimize, u and x−2u should be made as nearly equal as possible. Thus, we choose u = [x/3] or [x/3] + 1, whichever makes u and x−2u differ by no more than one, where [y] denotes the largest integer less than y.   The value of I(N) for any value of N can be found by direct iteration of the above equation. Values of I(x) for several values of x are given in Table 4.1.

Table 4.1.   Representative Values of I(x)

x	I(x)
2	1
3	1
4	2
7	2
9	2
17	3
27	3
3^k	k

INFINITE STAGE PROCESSES

Another class of problems that cannot be solved in a straight-forward fashion occurs when there is an explicit stage variable as in previous chapters, but when the number of stages becomes infinite. If the system equations and single-stage performance index do not depend on the value of the stage variable, as is the case in most problems of this type, then the procedures described in this section can be applied.

Let us attempt to minimize the criterion

$$J = \sum_{k=0}^{\infty} L(x(k),\ u(k)), \tag{4.4}$$

with system equation

$$x(k+1) = g(x(k),\ u(k)),\ x(0) = c, \tag{4.5}$$

and constraints

$$x \in X,\ u \in U\ .$$

If we set

$$I(x,k) = \min_{\substack{u(j) \\ j=k,\ldots,N}} \left\{ \sum_{j=k}^{N} L(x(j),\ u(j)) \right\}, \tag{4.6}$$

we obtain the relation

$$I(x,k) = \min_{u} [L(x,u) + I(g(x,u),\ k+1)],\ k=0,1,\ldots,N \tag{4.7}$$

Letting $N \to \infty$, we see that $I(x,k) = I(x,k+1)$, i.e., I does not depend on the stage and can be written as $I(x)$. Formally, we can then write

$$I(x) = \min_{u}[L(x,u) + I(g(x,u))] \tag{4.8}$$

Note that this relation, although derived from a problem with an
explicit stage variable, has the same general form as Eq. (4.2),
the basic equation for problems without an explicit stage variable.
As we shall see, some of the same conceptual ideas can be applied
here with a similar computational savings.

To establish existence and uniqueness for an equation such as
(4.8), hypotheses regarding the behavior of L and g are required.
One set of sufficient conditions for $I(x)$ to exist and remain
finite requires that (i) $L(x,u)$ is bounded for finite x and u
(ii) there exists a state $\bar{x}$ and control $\bar{u}$ such that $L(\bar{x},\bar{u}) = 0$,
and (iii) the specified state $\bar{x}$ can be reached from any admissible
state by applying a finite number of admissible controls. In most
cases where $I(x)$ remains finite these conditions are satisfied.

For those cases where $I(x)$ remains finite, two iterative
methods may be used to calculate $I(x)$ and the corresponding optimal
control policy $\hat{u}(x)$. The first method is called approximation in
function space. This method corresponds to the classical successive approximation technique. The method begins with a guess,
$I^0(x)$, for the true minimum cost function $I(x)$. A new value of
this function is obtained from

$$I^{(1)}(x) = \min_u \{L(x,u) + I^{(0)}(g(x,u))\} \qquad (4.9)$$

The corresponding policy, $\hat{u}^{(1)}(x)$, is the value of u for each x
that minimizes the right hand side of Eq. (4.9). The procedure
continues in this manner with $I^{(i+1)}(x)$ determined from $I^{(i)}(x)$
according to

$$I^{(i+1)}(x) = \min_u \{L(x,u) + I^{(i)}(g(x,u))\} \qquad (4.10)$$

and with $\hat{u}^{(i)}(x)$ determined as the value of u for each x that minimizes the right hand side of Eq. (4.10), until convergence is obtained.
Note that $I^{(i)}(x)$ is <u>not</u> a true minimum cost function, since

$I^{(i-1)}(x)$ is used on the right hand side of the equation in the basic iteration. For this reason, it is difficult to prove convergence to a global optimum. However, the method is easy to apply and quite fast; it performs reasonably well in problems with simple structure.

A more reliable procedure is approximation in policy space. This procedure begins with a guess at the optimal control policy, $\hat{u}^{(0)}(x)$. The cost corresponding to this policy is calculated by iterating

$$I^{(0,j+1)}(x) = L(x,\hat{u}^{(0)}(x)) + I^{(0,j)}(g(x,\hat{u}^{(0)}(x))) \qquad (4.11)$$

Note that this iteration requires no minimization. The initial value $I^{(0,0)}(x)$ is taken to be

$$I^{(0,0)}(x) = 0, \text{ for all } x. \qquad (4.12)$$

When $I^{(0)}(x)$ has been determined from this iteration, a new policy $\hat{u}^{(1)}(x)$ is determined as the value of u, for each x, that minimizes the right hand side of

$$I^*(x) = \min_{u} \{L(x,u) + I^{(0)}(g(x,u))\} \qquad (4.13)$$

Note that $I^*(x)$ is <u>not</u> $I^{(1)}(x)$, the true minimum cost function corresponding to $\hat{u}^{(1)}(x)$. This is because $I^{(0)}(g(x,u))$ is used on the right hand side of Eq. (4.13), not $I^{(1)}(g(x,u))$. It is thus necessary to repeat the procedure represented by Eq. (4.11) to obtain this quantity.

The combination of these two procedures continues at each iteration, with a minimum cost function $I^{(i)}(x)$, corresponding to the newly computed policy, $\hat{u}^{(i)}(x)$, being determined from

$$I^{(i,j+1)}(x) = L(x,\hat{u}^{(i)}(x)) + I^{(i,j)}(g(x,\hat{u}^{(i)}(x))) \qquad (4.14)$$

with

$$I^{(i,0)}(x) = 0 \ , \quad \text{all } x \tag{4.15}$$

and with a new optimal policy, $\hat{u}^{(i+1)}(x)$, being determined as the
value of u corresponding to each x that minimizes the right hand
side of

$$I^*(x) = \min_{u} \{L(x,u) + I^{(i)}(g(x,u))\} \tag{4.16}$$

EXAMPLE

4.3 Consider the problem with performance criterion

$$J = \sum_{k=0}^{\infty} [x^2(k) + u^2(k)]$$

and system equation

$$x(k+1) = x(k) + u(k) \ .$$

(a) Verify that the sufficient conditions for $\hat{u}(x)$ to exist and
remain finite are satisfied.

(b) Apply the approximation in function space technique to determine $I(x)$ and $\hat{u}(x)$. Use as initial guess $I^{(0)}(x) = 2x^2$.

(c) Apply the approximation in policy space technique to determine $I(x)$ and $\hat{u}(x)$. Use the initial policy $\hat{u}^{(0)}(x) = -x$.

(d) Compare the computational requirements and convergence speed
of the two methods.

(a) The sufficient condition can be checked as follows:

   (i) $(x^2 + u^2)$ is finite for all finite x and u.

   (ii) The state $\bar{x} = 0$ and control $\bar{u} = 0$ have the property that

   $$L(\bar{x},\bar{u}) = \bar{x}^2 + \bar{u}^2 = 0.$$

(iii)   The state x=0 can be reached from any other state in one step by applying the control $u^*(x^*) = -x^*$.

(b)   The iteration proceeds as follows:

$$I^{(1)}(x) = \min_{u} \{x^2 + u^2 + I^{(0)}(x+u)\}$$

$$= \min_{u} \{x^2 + u^2 + 2(x+u)^2\}$$

$$= x^2 + \left(-\frac{2}{3}x\right)^2 + 2\left(x - \frac{2}{3}x\right)^2$$

$$= \frac{5}{3}x^2 = 1.667\ x^2$$

The second iteration yields

$$I^{(2)}(x) = \min_{u} \{x^2 + u^2 + \frac{5}{3}(x+u)^2\}$$

$$= x^2 + \left(-\frac{5}{8}x\right)^2 + \frac{5}{3}\left(x - \frac{5}{8}x\right)^2$$

$$= \frac{13}{8}x^2 = 1.625\ x^2$$

The third iteration yields

$$I^{(3)}(x) = \min_{u} \{x^2 + u^2 + \frac{13}{8}(x+u)\}^2$$

$$= x^2 + \left(-\frac{13}{21}x\right)^2 + \frac{13}{8}\left(x - \frac{13}{21}x\right)^2$$

$$= \frac{34}{21}x^2 = 1.619\ x^2$$

After continued iteration, it is found that the steady-state value is reached at

$$I(x) = 1.618\ x^2$$

The corresponding optimal policy is

$$\hat{u}(x) = -0.618\ x$$

(c)   The iterations proceed as follows:

$$\hat{u}^{(0)}(x) = -x$$

The iterations on the minimum cost function yield

$$I^{(0,1)}(x) = x^2 + (-x)^2 + 0$$

$$= 2x^2$$

$$I^{(0,2)}(x) = x^2 + (-x)^2 + 2(x-x)^2$$

$$= 2x^2$$

$$I^{(0)}(x) = 2x^2$$

The next optimal policy becomes

$$\hat{u}^{(1)}(x) = \arg \min_{u} \{x^2 + u^2 + 2(x+u)^2\}$$

$$= -\frac{2}{3}x$$

The iterations on the minimum cost function yield

$$I^{(1,1)}(x) = x^2 + \left(-\frac{2}{3}x\right)^2 + 0$$

$$= \frac{13}{9}x^2 = 1.444 \ x^2$$

$$I^{(1,2)}(x) = x^2 + \left(-\frac{2}{3}x\right)^2 + \frac{13}{9}\left(x - \frac{2}{3}x\right)^2$$

$$= \frac{130}{81}x^2 = 1.605 \ x^2$$

Continued iterations yield

$$I^{(1)}(x) = 1.625 \ x^2$$

The next optimal policy becomes

$$\hat{u}^{(2)}(x) = \arg\min_{u}\{x^2 + u^2 + 1.625\ (x+u)^2\}$$

$$= -\frac{13}{21}\ x = -\ 0.619\ x$$

The iterations on the minimum cost function yield

$$I^{(2,0)}(x) = x^2 + \left(-\frac{13}{21}\ x\right)^2$$

$$= \frac{610}{441}\ x^2 = 1.383\ x^2$$

$$I^{(2,1)}(x) = x^2 + \left(-\frac{13}{21}\ x\right)^2 + \frac{610}{441}\left(x - \frac{13}{21}\ x\right)^2$$

Continued iterations yield

$$I^{(2)}(x) = 1.618\ x^2$$

The next optimal policy becomes

$$\hat{u}^{(3)}(x) = \arg\min_{u}\ \{x^2 + u^2 + 1.618\ (x+u)^2\}$$

$$= -\ 0.618\ x$$

The corresponding minimum cost function is

$$I^{(3)}(x) = 1.618\ x^2\ .$$

These latter two values are convergent values for $\hat{u}(x)$ and $I(x)$.

(d) It can be shown in this case that both methods converge to the true optimal value. The approximation in policy space method converges somewhat faster (for example, two iterations to get $I(x)$ correct to three decimal places, while approximation in function space takes four), but its computational requirements are substantially greater, primarily due to the additional iterations required to obtain the exact $I^{(i)}(x)$ corresponding to $\hat{u}^{(i)}(x)$.

EXAMPLE

4.4   Consider a discrete version of problem 4.3 with performance criterion

$$J = \sum_{k=0}^{\infty} [x^2(k) + u^2(k)],$$

system equation

$$x(k+1) = x(k) + u(k),$$

and constraints

$$0 \le x \le 5,$$
$$-5 \le u \le 0.$$

The quantization increments are taken to be $\Delta x=1$, $\Delta u=1$, so that $x = \{0,1,2,3,4,5\}$, $U = \{-5,-4,-3,-2,-1,0\}$.

(a)   Apply a discrete version of approximation in function space to obtain $I(x)$ and $u(x)$.   Use as initial guess $I^{(0)}(x) = 2x^2$.
(b)   Apply a discrete version of approximation in policy space to obtain $I(x)$ and $\hat{u}(x)$.   Use the initial policy $\hat{u}^{(0)}(x) = -x$.
(c)   Compare the computational requirements and convergence speeds of the two methods.
(d)   Compare the results with those of the continuous-state version, Problem 4.3.

(a)   The iterations proceed as follows:

$$I^{(0)}(x) = 2x^2$$

The first iteration requires solution of

$$I^{(1)}(x) = \min_{u \varepsilon U} \{x^2 + u^2 + I^{(0)}(x+u)\}$$

for $x \varepsilon X$.   The results can be obtained by solving an equivalent one-stage problem, where $I^{(0)}(x+u)$ is used as the "cost-to-go" at the next "stage".   Numerical results are summarized in Figure 4.6.

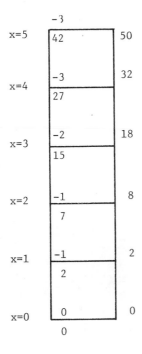

Figure 4.6   Solution for $I^{(1)}(x)$

The entries at grid points along the right-hand line of the figure are the $I^{(0)}(x)$ for $x \varepsilon X$.   The entries above and to the right of the grid points along the left-hand line of the figure are the $\hat{u}^{(1)}(x)$, $x \varepsilon X$, while the entries below and to the right are the $I^{(1)}(x)$, $x \varepsilon X$.

The next iteration requires solution of

$$I^{(2)}(x) = \min_{u \in U} \; \{x^2 + u^2 + I^{(1)}(x+u)\}$$

for $x \in X$. The procedure is similar to that of the previous iteration. Numerical results are shown in Figure 4.7.

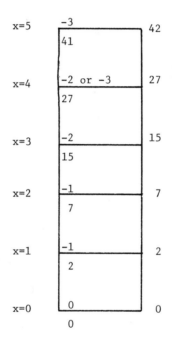

Figure 4.7   Solution for $I^{(2)}(x)$.

The entries at grid points along the right-hand line of the figure
are the $I^{(0)}(x)$ for $x \in X$. The entries above and to the right of
the grid points along the left-hand line of the figure are the
$u^{(1)}(x)$, $x \in X$, while the entries below and to the right are the
$I^{(1)}(x)$, $x \in X$.

Additional iterations do not change either $I^{(j)}(x)$ or $\hat{u}^{(j)}(x)$,
so the procedure has converged in two iterations.

(b) The iterations proceed as follows. We begin with

$$\hat{u}^{(0)}(x) = -x.$$

Iterations on the minimum cost function immediately yield

$$I^{(0,1)}(x) = x^2 + (-x)^2 + 0$$
$$= 2x^2$$

$$I^{(0,2)}(x) = x^2 + (-x)^2 + (x-x)^2$$
$$= 2x^2$$

Therefore,

$$I^{(0)}(x) = 2x^2$$

The next optimal policy is then obtained from

$$\hat{u}^{(1)}(x) = \arg\min_{u \varepsilon U} \{x^2 + u^2 + I^{(0)}(x+u)\}$$

for all x ε X. A procedure similar to that for part (a) can be used. Numerical results are shown in Figure 4.8. The entries at grid points along the right-hand line of the figure are the $I^{(0)}(x)$ for x ε X. The entries above and to the right of the grid points along the left-hand line of the figure are the $u^{(1)}(x)$, x ε X, while the entries below and to the right are the I(x), x ε X.

Further iterations do not change the optimal policy other than to show that $\hat{u}(4) = -2$ or -3, not just -3. Convergence is thus obtained in one iteration.

(c) As in the continuous case, both procedures obtain the same solution. Convergence is more rapid with approximation in policy space, although the computational burden per iteration is greater.

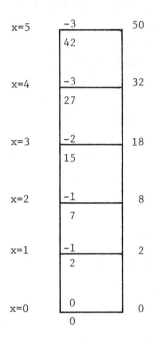

Figure 4.8   Solution for $\hat{u}^{(1)}(x)$

(d) A comparison of the discrete solution with the continuous
solution evaluated at quantized values of x is shown in Table 4.2.
Note that the solutions are generally in good agreement. The
largest discrepancy in the optimal control policy occurs at x=4,
where the continuous value of -2.472 is almost halfway between the
two values of -2 and -3 obtained in the discrete case. As would be
expected, the minimum costs are consistently slightly larger for
the discrete case than for the continuous case.

Table 4.2   Comparison of Continuous Solution and
Discrete Solution at Quantized States

	Continuous Case		Discrete Case	
x	u(x)	I(x)	u(x)	I(x)
5	-3.090	40.450	-3	42
4	-2.472	25.888	-2 or -3	27
3	-1.854	14.562	-2	15
2	-1.236	6.472	-1	7
1	-0.618	1.618	-1	2
0	0.000	0.000	0	0

FORWARD DYNAMIC PROGRAMMING

Although the recurrence relations and optimum decision sequence
recovery methods of the preceding chapters are appropriate
for most problems, there are certain cases where it is desirable to
use procedures in which the directions of the sweep in the stage
variable are reversed. Because the basic recursive relation is
operated forward in the stage variable for such procedures, these
methods are usually referred to as forward dynamic programming.
In what follows, we shall refer to the procedure of previous chapters
as backward dynamic programming.

There are two types of forward dynamic programming that have
received substantial attention. The more straightforward method is
simply a relabeling of the stage variable such that it represents
the number of stages to go, rather than the index of the actual
stage. If we simply substitute a new stage index ℓ where

$$\ell = N-k, \quad k=0,1,\ldots,N$$

then the basic recurrence relations take the form

$$I(x,\ell) = \min_{u} \{L(x,u,N-\ell) + I([g(x,u,N-\ell), \ \ell-1)]\}, \ell = 1,2,\ldots,N$$

(4.17)

$$I(x,0) = \min_{u} \{L(x,u,N)\} \hspace{3cm} (4.18)$$

Note that the stages at which we evaluate the functions $L(x,u,k)$ and $g(x,u,k)$ are given in terms of ℓ and the total number of stages N. Except for this slightly different way of evaluating stage-varying functions and the fact that the "stage" variable ℓ decreases rather than increases, the iterative method works exactly as in the backward case.

The above form of forward dynamic programming becomes more significant when we consider the stage-invariant case. If

$$L(x,u,k) = L(x,u) \quad \text{and} \quad g(x,u,k) = g(x,u),$$

then the equations become

$$I(x,\ell) = \min_{u} \{L(x,u) + I[g(x,u), \ \ell-1]\} \hspace{2cm} (4.19)$$

$$\ell = 1,2,\ldots,N$$

$$I(x,0) = \min_{u} \{L(x,u)\} \hspace{3cm} (4.20)$$

These equations can be interpreted as stating that the solution to a process with ℓ stages is given recursively in terms of the solution to an $(\ell-1)$ stage process. Under this interpretation, we are always starting from the initial stage, but we are allowing the length of the process to increase one stage at a time. Many derivations of the basic dynamic programming equation use precisely this viewpoint. Clearly, this interpretation is completely accurate for a stage-invariant system, but it does not extend straightforwardly to a stage-varying problem. For this reason we prefer the backward approach in our basic derivation.

There is another form of forward dynamic programming that is prevalent in the literature. In this case, a rather different form of the recurrence relation is obtained. This relation has certain computational properties that make it more useful than the backward approach for certain types of problems.

In this method, we define the minimum cost function differently; instead of I being the minimum cost that can be achieved starting at the present state and going to the end of the process, it is the minimum cost that can be achieved starting in an admissible initial state and arriving at the present state at the present time. If we write this minimum cost function as $I'(x,k)$ to distinguish it from the minimum cost function for the backward case, the definition becomes

$$I'(x,k) = \min_{u(0),u(1),\ldots,u(k-1)} \left\{ \sum_{j=0}^{k-1} L(x(j),u(j),j) \right\} \quad (4.21)$$

where

$$g(x(k-1), u(k-1), k-1) = x \quad (4.22)$$

In this equation, note that x is the last state along a trajectory, rather than the initial state as in the backward method.

The recurrence relation for this case is derived in a manner similar to the backward approach. First we define the function $g^{-1}[x(k+1),u(k),k]$ to be the state $x(k)$ from which the state $x(k+1)$ is reached at time k by applying the decision $u(k)$; formally

$$g[g^{-1}[x(k+1),u(k),k], u(k),k] = x(k+1) \quad (4.23)$$

Then, proceeding as before, we see that

$$I'(x,k) = \min_{u} \{L[g^{-1}(x,u,k-1),u,k-1] + I'[g^{-1}(x,u,k-1),k-1]\}$$

$$(4.24)$$

Note that this iterative procedure does in fact determine $I'(x,k)$ in terms of $I'(x,k-1)$; thus, the calculations do indeed proceed forward in the stage variable rather than backward.

In order to start the procedure, we must determine the minimum cost function at the initial state. If there is a specified initial state $x(0) = c$, then we define

$$I'(x,0) = 0, \quad x = c$$
$$= \infty, \quad x \neq c \qquad\qquad (4.25)$$

This initial condition may be implemented in a manner similar to that developed for a terminal constraint in the backward case, as discussed in Chapter 3. Alternatively, if the initial state is not specified but an initial cost function $\psi(x(0))$ is given, then we let

$$I'(x,0) = \psi(x) \qquad\qquad (4.26)$$

and proceed as before.

The corresponding optimum policy function $\hat{u}'(x,k)$ is a tabulation of the values of u for which the minimum in Eq. (4.24) is attained; formally,

$$\hat{u}'(x,k) = \arg\min_{u} L[g^{-1}(x,u,k-1),u,k-1] + I'[g^{-1}(x,u,k-1),k-1]$$

$$(4.27)$$

Note that this function tells us what decision should have been applied at stage $(k-1)$ in order for us to reach state x at stage k; again, it differs substantially from the optimum policy function for the backward case.

To solve the above recursive relation, we begin at the initial state and iterate forward until we obtain $I'(x,N)$ and $\hat{u}'(x,N)$. If the decision at stage N does not affect the performance criterion, i.e.,

if $L(x,u,N) = L(x,N)$,   the optimum terminal state is selected
by minimizing the sum of $I'(x,N)$ and $L(x,N)$, i.e., the optimum
terminal state $\hat{x}(N)$ is determined from

$$\hat{x}(N) = \arg \min_{x} [I'(x,N) + L(x,N)] \qquad (4.28)$$

A traceback is carried out by calculating $\hat{u}'(\hat{x}(N),N)$, determining
$\hat{x}(N-1)$ from

$$g^{-1}[\hat{x}(N),\hat{u}'(\hat{x}(N),N),N-1] , \qquad (4.29)$$

and continuing backward one stage at a time until $x(0)$ is reached.
If the decision at stage k does affect the performance criterion,
i.e., if $L(x,u,N)$ explicitly depends on u, then an additional stage
of iteration is necessary. The functions $I'(x,N+1)$ and $\hat{u}'(x,N+1)$
are calculated using $I'(x,N)$ and Eqs. (4.24) and (4.27). The
optimal state at stage (N+1) is calculated by minimizing $I'(x,N+1)$
over x, i.e.,

$$\hat{x}(N+1) = \arg \min_{x} [I'(x,N+1)] \qquad (4.30)$$

Note that this procedure is entirely equivalent to the method just
discussed except that an implicit terminal cost function of
$L(x,N+1) = 0$ has been assumed. The traceback then proceeds exactly
as before.

A very interesting question that arises in connection with
the forward dynamic programming method is the determination of conditions under which the same solution is obtained as for the normal
backward approach. Two results will be stated here without proof.

The first result is that for a given problem with a fixed
initial state and a fixed terminal state, both methods will obtain
exactly the same optimum decision sequences and optimal trajectories.
The second result is that for a given problem in which neither the
initial state nor the terminal state is specified, but the same
initial cost functions and terminal cost functions are given,

both methods will again determine exactly the same optimum
decision sequences and optimal trajectories, including the same
initial and final states. Clearly, many extensions of these
results can easily be obtained. Further extension and substantia-
tion of these results will be given in the Examples and Supplementary Problems of this and later chapters.

EXAMPLE

4.5   Consider the problem with cost function

$$J = \sum_{k=0}^{2} L(x(k),u(k),k)$$

where the functions $L(x(k),u(k),k)$, $k=0,1,2, \ldots$ are given by

$$L(x(0),u(0),0) = x^2(0) + u^2(0)$$
$$L(x(1),u(1),1) = x^2(1)$$
$$L(x(2),u(2),2) = x^2(2) + u^2(2)$$

The system equation is

$$x(k+1) = x(k) + u(k), \quad k=0,1,2$$

and the initial state is $x(0) = 1$. There are no constraints.

(a)   Find the minimum cost function and optimum decision policy
using the conventional backward approach. Find the optimum
decision sequences, optimal trajectory, and corresponding single-
stage costs from $x(0) = 1$.

(b)   Attempt to find the optimal decision sequence, optimal trajectory, and corresponding single-stage costs from $x(0) = 1$, using
the viewpoint that we first solve a one-stage problem recursively
in terms of the solution to the zero-stage problem and, then
solve a two-stage problem recursively in terms of the solution to
the one-stage problem. Does this answer differ from that obtained
in part (a)? Explain.

(c) Attempt to find the optimum decision sequence, optimal trajectory and corresponding single-stage costs from $x(0) = 1$ using the forward dynamic programming recursive equations in Eqs. (4.24) and (4.27). Does this answer differ from that obtained in part (a)? Explain.

(a) We apply the basic recurrence relations starting at stage 2. We obtain

$$I(x,2) = \min_u \{L(x,u,2)\} = \min_u \{x^2 + u^2\}$$

Clearly,

$$\hat{u}(x,2) = 0$$
$$I(x,2) = x^2$$

Next we solve

$$I(x,1) = \min_u \{L(x,u,1) + I(x + u,2)\}$$

$$= \min_u \{x^2 + (x+u)^2\}$$

We see that

$$\hat{u}(x,1) = -x$$
$$I(x,1) = x^2$$

Finally, we solve

$$I(x,0) = \min_u \{L(x,u,0) + I(x+u,1)\}$$

$$= \min_u \{x^2 + u^2 + (x+u)^2\}$$

The result is

$$\hat{u}(x,0) = -\frac{1}{2} x$$

$$I(x,0) = \frac{3}{2} x^2$$

The optimum decision sequence, optimal trajectory, and corresponding
single-stage costs from $x(0) = 1$ are given in Table 4.3.

Table 4.3   Optimum Decision Sequence, Optimal Trajectory,
             and Corresponding Single Stage Costs from
             $x(0) = 1$ Using Backward Approach

k	$\hat{x}(k)$	$\hat{u}$	$L(\hat{x}(k),\hat{u}(k),k)$
0	1.000	-0.500	1.250
1	0.500	-0.500	0.250
2	0.000	0.000	0.000
Total Cost	–	–	1.500

(b)   We first solve a zero-stage problem using $L(x(0),u(0),0)$ as
our criterion.

$$I(x,0) = \min_{u} \{L(x,u,0)\}$$

$$= \min_{u} \{x^2 + u^2\}$$

Clearly, the solution is

$$\hat{u}(x,0) = 0$$
$$I(x,0) = x^2$$

Assuming that this is the solution to a zero-stage problem, we
attempt to find a solution to a one-stage problem using

$$I(x,1) = \min_{u} \{L(x,u,0) + I(x+u,0)\}$$

$$= \min_{u} \{x^2 + u^2 + (x+u)^2\}$$

The result is

$$\hat{u}(x,1) = -\frac{1}{2} x$$

$$I(x,1) = \frac{3}{2} x^2$$

Assuming that this is the answer to a one-stage problem, we attempt
to find the solution to a two-stage problem using

$$I(x,2) = \min_{u} \{L(x,u,0) + I(x+u,1)\}$$

$$= \min_{u} \{x^2 + u^2 + \frac{3}{2} (x+u)^2\}$$

The result is

$$\hat{u}(x,2) = -\frac{3}{5} x$$

$$I(x,2) = \frac{8}{5} x^2$$

If we apply these results from $x(0) = 1$, we obtain the optimum
decision sequence, optimal trajectory, and corresponding single-
stage costs as shown in Table 4.4

Table 4.4   Optimum Decision Sequence, Optimal Trajectory,
             and Corresponding Single-Stage Costs from $x(0)=1$
             Using a Sequence of Progressively Longer-Stage
             Problems

k	$\hat{x}(k)$	$\hat{u}(k)$	$L(\hat{x}(k),\hat{u}(k),k)$
0	1.000	-0.600	1.360
1	0.400	-0.200	0.200
2	0.200	0.000	0.040
Total Cost			1.600

This solution differs from that obtained in part (a) because it
uses an incorrect recurrence relation in going from the zero-stage
problem to the one-stage problem. If the single-stage cost
function at stage 1 was $[x^2(1) + u^2(1)]$, then the single-stage costs
at each stage would have had the same form, $[x^2(k) + u^2(k)]$, and
the correct answer would have been obtained.

(c)   We note that the inverse function $g^{-1}(x(k+1),u(k),k)$ is given
by

$$g^{-1}(x(k+1),u(k),k) = x(k+1) - u(k)$$

The iterations then proceed in a straightforward fashion.   First,
we see that since the given initial stage is $x(1) = 0$,

$$I'(x,0) = 0 \qquad x = 1$$
$$= \infty \qquad x \neq 1$$

Then we solve for $I'(x,1)$ from

$$I'(x,1) = \min_{u} \{(x-u)^2 + u^2 + I'(x-u,0)\}$$

Clearly, the only allowable value for $x-u$ is 1, so that we obtain
directly

$$\hat{u}'(x,1) = x-1$$

$$I'(x,1) = (x-1)^2 + 1$$

We next calculate $I'(x,2)$ from

$$I'(x,2) = \min_{u} \{(x-u)^2 + I'(x-u,1)\}$$
$$= \min_{u} \{(x-u)^2 + (x-u-1)^2 + 1\}$$

We see that

$$\hat{u}'(x,2) = x - \frac{1}{2}$$

$$I'(x,2) = 1.5$$

Finally, we observe that in order to determine the state at $x=2$,
we must carry the procedure one step further and determine the
optimum state at $x=3$.   We evaluate $I'(x,3)$ from

$$I'(x,3) = \min_{u} \{(x-u)^2 + u^2 + I'(x-u,2)\}$$

$$= \min_{u} \{(x-u)^2 + u^2 + 1.5\}$$

We obtain

$$\hat{u}'(x,3) = \frac{1}{2} x$$

$$I'(x,3) = 1.5 + \frac{1}{2} x^2$$

Clearly, the minimum value of $I'(x,3)$ occurs at $x=0$, i.e., $\hat{x}(3)=0$. The corresponding previous decision is then $\hat{u}'(0,3) = 0$; in terms of our previous notation, this decision is $\hat{u}(2)$. We find the next point along the optimal trajectory as $g^{-1}(\hat{x}(3),\hat{u}(2),2) = 0$; this quantity is $\hat{x}(2)$. We then see that $\hat{u}'(0,2) = -\frac{1}{2} = \hat{u}(1)$ and that $g^{-1}(0, -\frac{1}{2}, 1) = \frac{1}{2} = \hat{x}(1)$. Finally, we obtain $\hat{u}'(\frac{1}{2},1) = -\frac{1}{2} = \hat{u}(0)$ and $g^{-1}(\frac{1}{2}, -\frac{1}{2}, 0) = 1 = \hat{x}(0)$. These results are summarized in Table 4.5.

Table 4.5   Optimum Decision Sequence, Optimal Trajectory and Corresponding Single-Stage Costs from $x(0)=1$ Using the True Forward Dynamic Programming Approach

k	$\hat{x}(k)$	$\hat{u}(k)$	$L(\hat{x}(k),\hat{u}(k),k)$
0	1.000	-0.500	1.250
1	0.500	-0.500	0.250
2	0.000	0.000	0.000
3	0.000	-	-
Total Cost			1.500

These are the same results as obtained in part (a). Note that $\hat{x}(3)$ as obtained here is indeed the next state if we use the decision policy from part (a) on the state $x(2)$ obtained in part (a), i.e.,

$$\hat{x}(3) = \hat{x}(2) + \hat{u}(\hat{x}(2),2) = 0 + 0 = 0$$

These results are identical to part (a) because we use the same initial state, $x(0) = 1$, and because we have the same implicit terminal cost function at stage 3, namely $L(x,u,3) = 0$.

A computational procedure for the discrete-state case, similar to that for backward dynamic programming in Chapter 3, can be developed for forward dynamic programming. This procedure can be stated as follows: at a given quantized x and k, try all possible quantized decisions $u(k-1)$; find the corresponding previous state $x(k-1)$; evaluate the quantity inside the brackets in Eq. (4.24), using interpolation procedures if necessary; and pick the optimal decision and minimum cost.

This procedure has some deficiencies. In the first place there is no a priori way of ensuring that a given decision $u(k-1)$ is in fact an admissible decision at $x(k-1) = g^{-1}(x,u(k-1),k)$. Second, it may be difficult to compute the functional g^{-1} if g is a nonlinear, time-varying functional.

A computational procedure which overcomes these difficulties and offers other advantages as well is the following: at each quantized admissible decision, for each corresponding next state, $x(k) = g[x(k-1), u(k-1), k-1]$, check to see if it has been the next state for any decision applied at previous values of $x(k-1)$; if it has not previously been a next state, then store the quantity in braces in Eq. (4.24) as the tentative minimum cost at that point; if it has been the next state, compare the quantity in braces in Eq. (4.24) with the tentative minimum cost already computed at that point, and if it is less than this minimum cost, replace the value stored there.   This procedure continues until the quantized admissible decisions have been applied at every quantized state x(k-1). The tentative minimum costs and optimal decisions at each x(k) = g[x(k-1),k-1] are then the true minimum costs and optimal decisions at these points.

This procedure is seen more clearly in a simple one-dimensional example. The system equation for this problem is

$$x(k+1) = x(k) + u(k).$$

The performance criterion is

$$J = \sum_{j=0}^{4} [x^2(j) + u^2(j)]$$

The constraints are

$$-1 \leq u(k) \leq 1 ,$$

and

$$0 \leq x(k) \leq 2 .$$

The initial state is known to be
$$x(0) = 2$$

The quantization increments are taken to be $\Delta x=1$ and $\Delta u=1$. The first step in the procedure is to apply the three quantized admissible decisions, u = -1, u = 0, and u = +1, at the initial state x(0) = 2. The minimum cost and optimal decision for admissible states at x(1) are found without need for comparison. These values are shown in Figure 4.9, where the minimum cost appears above and the optimal decision below the quantized state and stage to which these quantities correspond. Note that the optimal decision determines what state to have come from, not what state to go to.

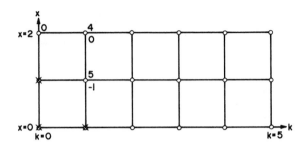

Figure 4.9   First Step in Forward Dynamic
Programming Procedure

In going from I'(x,1) to I'(x,2) a comparison is required in
some of the calculations. If the decisions are first applied at
x(1)=2, the tentative minimum costs and optimal decisions are shown
in Figure 4.10. Asterisks are placed beside these values to show
that they are tentative.

When the controls at x(1) are applied, the states x(3) = 2 and
x(2) = 1 are possible next states. The minimum costs coming from
x(1) = 1 are compared with the values already there in Figure 4.8.
In both cases, less cost is obtained when x(1) = 1. The complete
results at k=2 are as shown in Figure 4.11.

This procedure continues until the minimum cost and optimal
control have been computed at all quantized values of x and k. The
results are shown in Figure 4.12.

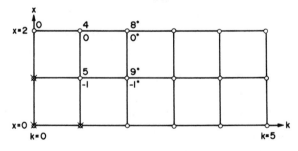

Figure 4.10   Tentative Minimum Costs at k=2

EXTENSIONS OF THE BASIC PROCEDURES247

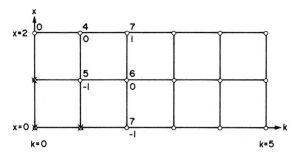

Figure 4.11   Complete Results at k=2

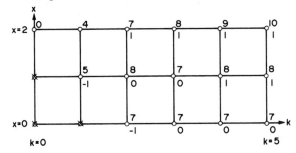

Figure 4.12   Complete Results for Forward Dynamic
Programming Example

These results can be used in a number of different ways. First
suppose that the final state can be anywhere in the region of
$0 \leq x \leq 2$. In this case a second search is made over the minimum
costs at the quantized final states, and the final state is taken
to be the one for which the minimum cost is smallest. In this example,
the minimum cost is $I' = 7$ at $x(5) = 0$. The optimal trajectory is
then as shown in Figure 4.13.

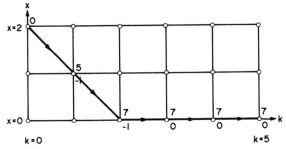

Figure 4.13   Optimal Trajectory if Final State is Not
Constrained

Next, suppose that a terminal cost function is added at k=5.
Let this terminal cost function be

$$\psi[x(5), 5] = 2.5 [x(5)-2]^2$$

This function can then be added to the minimum costs $I'[x(5),5]$
shown in Figure 4.12. The resulting total cost is obtained for
x=0, 1 and 2 in Table 4.6. The minimum total cost is thus seen to
be 10, corresponding to x(5) = 2. The optimal trajectory corresponding to this final state is as shown in Figure 4.14.

Table 4.6   Total Cost in Forward Dynamic
             Programming Example

x	$I'(x,5)$	$\psi(x,5)$	Total Cost
2	10	0	10
1	8	2.5	10.5
0	7	10	17

The same problem with the same terminal cost function was
solved by backward dynamic programming in Chapter 3. From the
results presented there, it can be seen that for the initial state
x=2, the minimum cost and optimal trajectory are exactly the same
as shown in Figure 4.14. As noted earlier, for the same problem
with the same initial and final states, forward dynamic programming

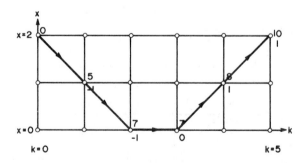

Figure 4.14   Optimal Trajectory corresponding to x(5)=3
               (Optimal Trajectory for the Given Terminal
               Cost Function)

and backward dynamic programming obtain the same minimum cost and
optimal trajectory. Of course, the costs and controls at intermediate states are completely different; this is due to the difference
in definition between $I(x,k)$ and $\hat{u}(x,k)$ and $I'(x,k)$ and $\hat{u}'(x,k)$.

This example illustrates a number of the useful properties of
the forward dynamic programming solution. In the first place,
there is great flexibility in the terminal cost function and/or
terminal constraints that can be applied. In particular, a terminal
cost function can be added after all the computations have been
made, and the terminal state that minimizes the total cost can then
be selected. It is thus possible to assess the effects of using a
number of different terminal cost functions without repeating the
forward dynamic programming calculations. On the other hand, in
backward dynamic programming it is necessary to repeat the entire
computation for each different terminal cost function.

Using the forward dynamic programming solution it is easy to
examine trajectories reaching many different terminal states; thus,
if an explicit terminal cost function is not known, the trajectories
corresponding to several of the lowest values of minimum cost can
be examined and the optimal trajectory selected on the basis of
more subjective grounds, such as simplicity of realization, reliability, convenience to humans associated with the system, etc. If a
trajectory other than the one corresponding to the lowest cost is
chosen, then the added cost in terms of the quantities in the performance criterion is given explicitly.

The application of terminal constraints can be done in exactly
the same way with similar results.

Another useful property of forward dynamic programming is
that the initial state can easily be constrained to one state,
while with backward dynamic programming optimal trajectories for
all admissible states at the initial time are found. This is true

even when in backward dynamic programming the computations at k=0
are made for only one initial state: in this case computations have
still been done at a number of points through which no trajectory
from the initial state will pass. Thus, when it is desired to have
an optimal trajectory from a given initial state, a saving in computing time can be achieved by using forward dynamic programming.

The extension of this forward dynamic programming procedure to
problems for which the next states $x(k) = g[g(k-1),u(k-1)]$ do not
occur exactly at quantized values is straightforward. This can be
done in a particularly effective way by associating the next state
with the nearest quantized state and interpolating in the n-dimensional state space to compare the cost in brackets in Eq. (4.24)
with the tentative minimum cost at the quantized state. Thus, the
interpolations can be used for comparison purposes only, and the
minimum cost and optimal control can be evaluated exactly along a
true trajectory. This eliminates the need for interpolations in
reconstructing optimal trajectories after the computations have
been completed. The only purpose which quantization in the state
variables serves is to divide the state space into equivalence classes
over which the minimum costs are compared. The minimum cost and
optimal control associated with a given quantized state are not
necessarily evaluated exactly at the quantized state but can be at
any state within this equivalence class.

One disadvantage of forward dynamic programming is that the
feedback control property of backward dynamic programming is not
retained. The optimal controls determine what the previous state
should have been, rather than the next state. Thus, if a deviation
occurs from the selected optimal trajectory, a new optimal trajectory
cannot be easily found.

The type of problem to which forward dynamic programming is
best suited is one in which the initial state is specified, flexibility in choosing the terminal state is desirable, and computation

of a new optimal trajectory is feasible if deviations occur from the original trajectory. One class of problems where these conditions are met is on-line control and dispatching applications. Another class of such problems come about when it is desired to pre-compute the response of a system which always starts from a known initial state to a large number of possible conditions. Still another class of problems is optimum system planning; in these cases the present system configuration is known and flexibility in assessing the results of different terminal cost functions is often important.

EXAMPLE

4.6   Consider the problem with performance criterion

$$J = \sum_{k=0}^{3} \left(\frac{2 + u(k)}{1 + x^2(k)} \right) \quad ,$$

system equations

$$x(k+1) = x(k) + [2 - \frac{3}{2} x (k) + \frac{1}{2} x^2(k)] \, u(k),$$

and constraints

$$0 \leq x \leq 2 \; ,$$
$$-1 \leq u \leq 1 \; .$$

We wish to apply the basic forward dynamic programming computational procedure to this problem using uniform quantization increments of $\Delta x = 1$ and $\Delta u = 1$.

(a)   Obtain the complete solution for stages k = 0 to k = 4.

(b)   Determine the optimal state trajectory and decision sequence if the terminal state can be any quantized admissible state at stage 4.

(c)   Determine the optimal state trajectory and decision sequence if there is a terminal penalty at stage k = 4 of $\phi(x,4) = \frac{1}{2} x^2$.

(d) Determine the optimal state trajectory and decision sequence
if the terminal state can occur at <u>either</u> stage 3 or stage 4 and
if there is a terminal penalty at stage 3 of

$$\phi(x,3) = 0.2 + 0.3 \ (x-2)^2,$$

and at stage 4 of

$$\phi(x,4) = 0.1 \ (x-2)^2$$

(a) Although the system equation and single-stage cost function
are complex functions, they do not vary with the stage. Therefore,
it is possible to precompute these functions for the allowed values
of x and u and retrieve the appropriate values where needed. The
next state as a function of current state and current decision is
given in Table 4.7.

Table 4.7   Next State Table

x(k) \ u(k)	-1	0	+1
2	1	2	3
1	0	1	2
0	-2	0	2

The single-stage cost function as a function of current state and
current decision is shown in Table 4.8

Table 4.8   Single-Stage Cost Table

x(k) \ u(k)	-1	0	1
2	0.2	0.4	0.6
1	0.5	1.0	1.5
0	1.0	2.0	3.0

Using these tables, the procedure can be implemented in a straightforward fashion. The results of the calculations are shown in
Figure 4.16, where the number above and to the right of the grid
point is the minimum cost and the number below and to the right of
the grid point is the optimum previous decision

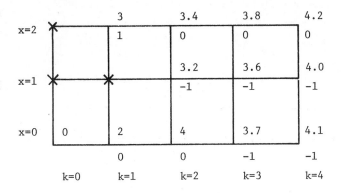

Figure 4.16   Complete Solution

(b) If the final state occurs at stage 4, but is otherwise free,
then a comparison over the minimum costs at these states is required.
From examination of Figure 4.16, it is clear that the minimum cost
is 4.0, corresponding to x=1. The state trajectory and optimum
decision sequence are as shown in Table 4.9.

Table 4.9   Optimum State Trajectory and Decision
Sequence for $x(4) = 1$

k	x(k)	u(k)	L(x,u,k)
0	0	1	3.0
1	2	0	0.4
2	2	0	0.4
3	2	-1	0.2
4	1	--	--
Total			4.0

(c)   If there is a terminal cost at stage 4 of

$$\phi(x,4) = \frac{1}{2} x^2,$$

then there is a trade-off between costs prior to stage k=4 and
the terminal costs at stage k = 4.   The sum of these two costs is
shown in Table 4.10.   The minimum is clearly attained at x=0.   The
corresponding state trajectory and optimum decision sequence are
shown in Table 4.11.

Table 4.10   Total Cost at Stage k=4 Using the Terminal
Cost Function $\phi(x,4) = 1/2 \, x^2$

x	$I'(x,4)$	$\phi(x,4)$	Sum
2	4.2	2.0	6.2
1	4.0	0.5	4.5
0	4.1	0.0	4.1

Table 4.11   Optimum State Trajectory and Decision
Sequence with $\phi(x,4) = 1/2 \, x^2$

k	$\hat{x}(k)$	$\hat{u}(k)$	$L(\hat{x}(k),\hat{u}(k),k)$
0	0	1	3.0
1	2	0	0.4
2	2	-1	0.2
3	1	-1	0.5
4	0	--	0.0
Total			4.1

(d)   If the terminal cost at stage 3 is

$$\phi(x,3) = 0.2 + 0.3 \, (x-2)^2$$

and if the terminal cost at stage 4 is

$$\phi(x,4) = 0.1 \, (x-2)^2,$$

then the total cost for all states at stages 3 and 4 is as shown
in Table 4.12

Table 4.12   Total Cost at Stages 3 and 4 Using the
Terminal Cost $\phi(x,3) = 0.2 + 0.3 \ (x-2)^2$
and $\phi(x,4) = 0.1 \ (x-2)^2$

k	x	I'(x,k)	$\phi(x,k)$	Sum
3	2	3.8	0.2	4.0
3	1	3.6	0.5	4.1
3	0	3.7	1.4	5.1
4	2	4.2	0.0	4.2
4	1	4.0	0.1	4.1
4	0	4.1	0.4	4.2

The minimum is clearly attained at $x(3) = 2$. The corresponding
state trajectory and decision sequence is shown in Table 4.13.

Table 4.13   Optimum State Trajectory and Decision
Sequence for $\phi(x,3) = 0.2 + 0.3 \ (x-2)^2$
and $\phi(x,4) = 0.1 \ (x-2)^2$

k	$\hat{x}(k)$	$\hat{u}(k)$	$L(\hat{x}(k),\hat{u}(k)k)$
0	0	1	3.0
1	2	0	0.4
2	2	0	0.4
3	2	--	0.2
4	--	--	--
Total			4.0

SOLVED PROBLEMS

4.1 Consider the natural gas pipeline network pictured in Figure
4.17. The operation of the compressor stations A-L is such that
the flow within a given link is in the direction indicated by the
arrow. The capacity of each link is as shown on the figure. The
capacity of two links in a series is the smaller of the two
capacities. Find the path through the network from A to L with
the maximum capacity. Specify the nodes corresponding to the
compressor stations along the optimal path.

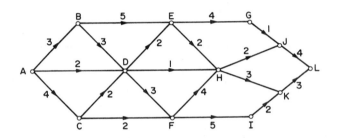

Figure 4.17   Gas Pipeline Network

     This problem is very similar to Example 4.1 in the Chapter,
except that the cost function is not a straightforward summation
of the costs along each link. Instead, the performance criterion
is related to maximization of the minimum capacity along the path.
Thus the problem not only has nonstandard system equations, but a
non-standard performance criterion.

     As indicated in Chapter 2, this criterion can be handled
easily within the basic framework of dynamic programming. We can
use the result developed there for the min-max case to show that

$$I(x,k) = \max_{u} \ \{\min [L(x,u)], \ \{I[g(x,u)]\}\}$$

In order to apply this equation, the system equation is
interpreted as in Example 4.1, i.e., x is the index of the present
node, u is the index of a link emanating from that node, and g(x,u)
specifies the node at the terminal end of this link. The function
L(x,u) is taken to be the capacity of the link. The procedure
then can be carried out much as in the previous problem, starting
with the terminal boundary conditions $I(L) = \infty$, i.e., there is
infinite capacity of the link from L, the terminal node of the network, to itself.

The solution for this particular network is shown pictorially
in Figure 4.18. The capacity of the path to the terminal node
from each intermediate node is shown beside the node, and the first
link along this path is shown by the directed arrow leaving the
node. The capacity of the optimal path from node A to node L is 3;
it is seen by tracing out this path that it goes from node A to
node B to node D to node F to node H to node K to node L.

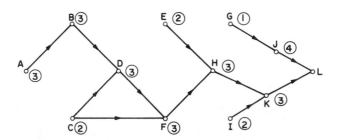

Figure 4.18   Solution to Example 4.1

4.2 A man is standing in a queue waiting for service, with N people
ahead of him. If he waits out the queue he receives a return of r.
On the other hand, he incurs a cost of c per unit time for waiting
in the queue. If he leaves the queue at any time without service
he obtains no return. If he knows that the probability that a
person will be served in a unit of time is p, determine his waiting
policy if he wishes to maximize his expected return.

As in the previous problem, it is not necessary to define a
stage variable explicitly. Let I(x) denote the expected return
obtained when there are x people ahead and the man employs an
optimal waiting policy. At each unit of time the man has the choice
of either waiting one more unit of time or leaving the queue. If
he waits he receives the expected return [pI(x-1) + (1-p)I(x) - c],
and if he leaves he obtains 0. Applying the Principle of Optimality,
we see that I(x) is given by

$$I(x) = \max \begin{cases} W: & [pI(x-1) + (1-p)\,I(x) - c] \\ \\ L: & 0 \end{cases}, \; x=1,2,\ldots,$$

where W denotes the decision to wait and L to leave. Clearly,

$$I(0) = r$$

The above equation is obviously equivalent to

$$I(x) = \max \begin{cases} W: & I(x-1) - \dfrac{c}{p} \\ \\ L: & 0 \end{cases}$$

To determine the optimal policy and the sequence of functions I(x),
we see that

$$x = 0: \quad I(0) = r > 0, \text{ optimal policy is } W$$

$$x = 1: \quad I(1) = \max \begin{cases} W: & r - \dfrac{c}{p} \\ L: & 0 \end{cases}, \text{ policy is } \begin{cases} W \text{ if } pr - c > 0 \\ L \text{ if } pr - c \le 0 \end{cases}$$

$$x = 2: \quad I(2) = \max \begin{cases} W: & r - \dfrac{2c}{p} \\ L: & 0 \end{cases}, \text{ policy is } \begin{cases} W \text{ if } pr - 2c > 0 \\ L \text{ if } pr - 2c \le 0 \end{cases}$$

In general, it is found that

$$I(x) = \max \begin{cases} W: & r - \dfrac{xc}{p} \\ L: & 0 \end{cases}, \text{ policy is } \begin{cases} W \text{ if } pr - xc > 0 \\ L \text{ if } pr - xc \le 0 \end{cases}$$

4.3   A perfectionist bartender working in a new place wants to mix
a large batch of martinis.   He has a quart bottle of gin (1 quart
= 32 oz.), and he wishes to make martinis that are exactly 8-1.
Therefore, he needs 4 oz. of vermouth.   However, the only measuring
glasses he can find hold 3 oz., and 5 oz., respectively.   After some
thought (he is also a mathematician), he finds that he can obtain
exactly 4 oz. by a combination of filling glasses, emptying glasses,
and pouring the contents of one glass into the other.   Find the
minimum number of pours required to obtain 4 oz.
(a) using a dynamic programming formulation with an explicit
stage variable; and
(b)   treating the problem as finitely terminable problem.

(a)   The state variables for the problem are the amounts of vermouth in each measuring glass.   We define two state variables by

x_1 = amount of vermouth in 3-oz. glass

x_2 = amount of vermouth in 5-oz. glass.

The state vector $\underline{x}$ contains these two variables, i.e.,

$$\underline{x} = \begin{bmatrix} x_1 \\ x_2 \end{bmatrix}$$

Then we see that the capacities of the glasses introduce the
following constraints

$$0 \le x_1 \le 3,$$

$$0 \le x_2 \le 5,$$

Let us use a quantization increment of one ounce for each state
variable, i.e., $\Delta x_1 = \Delta x_2 = 1$.   The set of admissible quantized
states then becomes

$$X = \left\{ \begin{bmatrix} x_1 \\ x_2 \end{bmatrix} , \quad x_1 = 0,1,2,3, \quad x_2 = 0,1,2,3,4,5 \right\}$$

$$= \left\{ \begin{bmatrix} 0 \\ 0 \end{bmatrix} , \begin{bmatrix} 0 \\ 1 \end{bmatrix} , \begin{bmatrix} 0 \\ 2 \end{bmatrix} , \begin{bmatrix} 0 \\ 3 \end{bmatrix} , \begin{bmatrix} 0 \\ 4 \end{bmatrix} , \begin{bmatrix} 0 \\ 5 \end{bmatrix} , \right.$$

$$\begin{bmatrix} 1 \\ 0 \end{bmatrix} , \begin{bmatrix} 1 \\ 1 \end{bmatrix} , \begin{bmatrix} 1 \\ 2 \end{bmatrix} , \begin{bmatrix} 1 \\ 3 \end{bmatrix} , \begin{bmatrix} 1 \\ 4 \end{bmatrix} , \begin{bmatrix} 1 \\ 5 \end{bmatrix} ,$$

$$\begin{bmatrix} 2 \\ 0 \end{bmatrix} , \begin{bmatrix} 2 \\ 1 \end{bmatrix} , \begin{bmatrix} 2 \\ 2 \end{bmatrix} , \begin{bmatrix} 2 \\ 3 \end{bmatrix} , \begin{bmatrix} 2 \\ 4 \end{bmatrix} , \begin{bmatrix} 2 \\ 5 \end{bmatrix} ,$$

$$\left. \begin{bmatrix} 3 \\ 0 \end{bmatrix} , \begin{bmatrix} 3 \\ 1 \end{bmatrix} , \begin{bmatrix} 3 \\ 2 \end{bmatrix} , \begin{bmatrix} 3 \\ 3 \end{bmatrix} , \begin{bmatrix} 3 \\ 4 \end{bmatrix} , \begin{bmatrix} 3 \\ 5 \end{bmatrix} \right\}$$

We next characterize the decision variables (fill or pour) by
how they affect the state variables (amount in glass). We define
two decision variables as

u_1 = amount of vermouth poured into 3-oz. glass

u_2 = amount of vermouth poured into 5-oz. glass

The decision vector $\underline{u}$ contains these two variables, i.e.,

$$\underline{u} = \begin{bmatrix} u_1 \\ u_2 \end{bmatrix}$$

We can enumerate the eight possible alternative actions we have at
any time. They are:
(1) fill 3-oz. glass
(2) fill 5-oz. glass
(3) empty 3-oz. glass
(4) empty 5-oz. glass

(5) pour from 3-oz. glass to 5-oz. glass until 5-oz. glass is filled

(6) pour from 3-oz. glass to 5-oz. glass until 3-oz. glass is empty

(7) pour from 5-oz. glass to 3-oz. glass until 3-oz. glass is filled

(8) pour from 5-oz. glass to 3.0z. glass until 5-oz. glass is empty.

Values of u_1 and u_2 corresponding to each alternative are then defined in terms of the current contents of the glasses.

For example, the first alternative, fill the 3-oz. glass, corresponds to $u_1 = 3 - x_1$, $u_2 = 0$. The complete set of alternatives, decisions, in the order (1)-(8) above, can be written

$$U = \left\{ \begin{bmatrix} 3 - x_1 \\ 0 \end{bmatrix}, \begin{bmatrix} 0 \\ 5 - x_2 \end{bmatrix}, \begin{bmatrix} -x_1 \\ 0 \end{bmatrix}, \begin{bmatrix} 0 \\ -x_2 \end{bmatrix}, \right.$$

$$\left. \begin{bmatrix} -(5-x_2) \\ 5-x_2 \end{bmatrix}, \begin{bmatrix} -x_1 \\ x_1 \end{bmatrix}, \begin{bmatrix} 3 - x_1 \\ -(3-x_1) \end{bmatrix}, \begin{bmatrix} x_2 \\ -x_2 \end{bmatrix} \right\}$$

The system equations then become

$$x_1(k+1) = x_1(k) + u_1(k)$$

$$x_2(k+1) = x_2(k) + u_2(k)$$

In vector notation, the equation is

$$\underline{x}(k+1) = \underline{x}(k) + \underline{u}(k)$$

The performance criterion is the minimum number of operations in the set U required to transform the state from $\underline{x}^T = (0,0)$ to either $\underline{x}^T = (0,4)$, $(1,4)$, $(2,4)$, or $(3,4)$, i.e., to one of the states where $x_2 = 4$. This can be expressed mathematically as

$$J = \sum_{k=0}^{N-1} 1$$

where N is the first stage at which $x_2 = 4$. The initial state is

$$\underline{x} = \begin{bmatrix} 0 \\ 0 \end{bmatrix}.$$

We now define $I(\underline{x},k)$ for all $x \in X$ as the minimum number of operations required to transfer the current state $\underline{x}$ to one of the four admissible terminal states where $x_2 = 4$. We obtain the dynamic programming equation as

$$I(\underline{x},k) = \min_{\underline{u} \in U} \{1 + I(\underline{x} + \underline{u},\ k{+}1)\}$$

If the set of four admissible terminal states is denoted as X^*, we obtain the terminal condition as

$$I(\underline{x},N) = 0, \quad \underline{x} \in X^*$$

$$= \text{undefined otherwise}$$

The value of N is carried as a running index and determined by defining $k=0$ at the first stage for which one can reach one of the states in X^*.

To obtain numerical results, we define a grid of admissible states as in Figure 4.19. For each stage k, $0 \le k \le K$, we mark with an $\times$ each state for which we can reach one of the allowed terminal states, $\underline{x} \in X^T*$, by stage N. We shall call such states terminable states. We also show by an arrow the first transition from this state required to achieve termination, i.e., we draw an arrow from this state to the resulting next state if this first transition is applied. We start at stage N and work back until 0, i.e., until

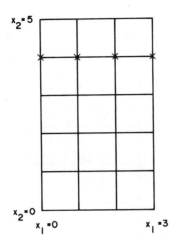

Figure 4.19   Set of Terminable States at k=N

there is an x in state $\underline{x}^T = (0,0)$. At stage N the terminable states
are the four admissible terminal states, as shown in Figure 4.19.
At stage N-1 we find two new terminable states, as shown in Figure
4.20

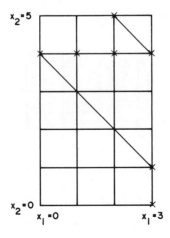

Figure 4.20   Set of Terminable States at k=N-1

At stage N-2 we find six more terminable states. The transitions from these new states, as well as the complete set of terminable states, are shown in Figure 4.21.

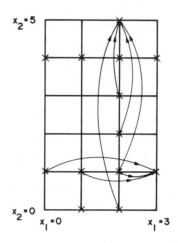

Figure 4.21   Set of Terminable States at k=N-2

If we continue this process, we find that the new terminable states
for k = N-6 are as shown in Figure 4.22.

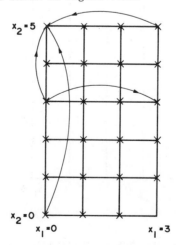

Figure 4.22   Terminable States at k = N-6 = 0

Note that one of the new states is $\underline{x} = \begin{bmatrix} 0 \\ 0 \end{bmatrix}$. Therefore, we have

arrived at stage 0 and, since N-6=0, we see that N=6, i.e., the
number of operations required is 6. By tracing forward the indicated transitions one stage at a time, we find that the optimum
sequence is

1. Fill 5-oz. glass, $\underline{x}^T = (0,5)$
2. Pour contents of 5-oz. glass into 3-oz. glass, $\underline{x}^T = (3,2)$
3. Empty 3-oz. glass, $\underline{x}^T = (0,2)$
4. Pour contents of 5-oz. glass into 3-oz. glass, $\underline{x}^T = (2,0)$
5. Fill 5-oz. glass, $\underline{x}^T = (2,5)$
6. Pour contents of 5-oz. glass into 3-oz. glass, $\underline{x}^T = (3,4)$

(b) The problem can be solved more compactly by following the
procedure for finitely terminable systems. To begin this procedure,
enter a zero at each grid point corresponding to the four admissible
terminable states. From each of the other grid points try all
possible decisions and enter a one if there is a decision that
results in the next state having a zero beside it. For such a
grid point keep a record of the decisions that achieve this.
Continue this procedure by applying all possible decisions from
any grid point not having a number beside it, and entering k+1
beside a grid point the first time that there is a decision
resulting in a next state having a k beside it. During this procedure, tabulate for each grid point the decision that achieves
this result. When this procedure has been carried out until all
grid points have a number beside them, the minimum number of operations to reach one of the terminal states from any possible initial
state is given by the number beside the grid point corresponding to
the initial state. The optimum sequence from any initial state can
be found by working forward from the decisions tabulated for each
state during the procedure. The result is shown in Figure 4.23.
The same optimum strategy as determined in part (a) is obtained.

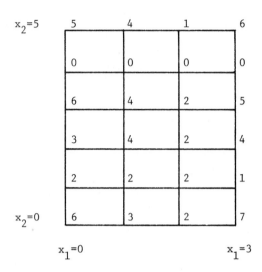

Figure 4.23   Grid of Minimum Number of Operations
               Required

4.4   Let us consider an important problem from the field of aeronautical engineering, namely that of minimum-time-to-climb for an
aircraft.   For each of the two problems discussed below, develop a
solution procedure based on dynamic programming and carry out the
solution.   The approach developed for finitely terminable systems
will be helpful here.

(a)   From wind-tunnel data for an experimental aircraft, it is
found that the time to climb 15,000 feet at constant velocity is
given as Δt_c in Table 4.14.   The time to accelerate 0.5 M at
constant altitude is given as Δt_a in Table 4.15.   If only these
climb and accelerate maneuvers are allowed, find the minimum time
and corresponding sequence of maneuvers to reach h = 75,000 ft,
and v = 3.0 M from h = 0, v = 0.5 M.

(b)   In addition to the climb and accelerate maneuvers described
in part (a), assume that the aircraft is also allowed to "zoom",
where a zoom is defined to be a dive of 15,000 ft. during which the

aircraft increases its speed by 0.5M. If the time to complete a
zoom is given by Δt_z in Table 4.16, recompute a new minimum time
and corresponding sequence of maneuvers to reach h = 75,000 ft. and
v = 3.0M from h = 0, v = 0.5 M.

Table 4.14 Δt_c (h,v) = Time to Climb 15,000 ft. at
              Constant Velocity from Altitude h at
              Velocity v

h in ft.

60,000	16	12	12	12	12	11
45,000	16	14	13	13	12	11
30,000	18	17	13	14	16	16
15,000	22	20	19	19	20	22
0	25	25	24	22	25	28

       0.5   1.0   1.5   2.0   2.5   3.0

               v in Mach no.

Table 4.15 Δt_a (h,v) = Time to Accelerate 0.5M at
              Constant Altitude from Velocity at
              Altitude h

h in ft.

75,000	18	14	15	16	17
60,000	18	16	16	18	19
45,000	19	17	18	19	20
30,000	20	18	19	20	21
15,000	21	20	18	19	21
0	22	22	22	27	30

       0.5   1.0   1.5   2.0   2.5

               v in Mach no.

Table 4.16 $\Delta t_z(h,v)$ = Time to Dive 15,000 Ft. and
              Accelerate 0.5 M from Altitude h and
              Velocity v

h in ft

75,000	7	6	5	4.5	4.5
60,000	7	6	5	4.5	4.5
45,000	8	7	6	5	4.5
30,000	10	8	7	6	5
	0.5	1.0	1.5	2.0	2.5

v in Mach no.

(a) By noting the similarity of this problem to problem 3.15, we
see that there are two state variables, altitude h and velocity
v, and two control variables, change in altitude and change in
velocity. We could easily set up all the formalism of problem 3.15
and solve it in exactly the same fashion. However, let us instead
attempt to utilize the concepts of finitely terminable systems to
reduce the computations.

First, observe that at altitude h = 75,000 ft. there is only
one allowable maneuver, namely accelerate. Similarly, for velocity
v = 3.0 M, there is only one allowable maneuver, namely, climb.
Since these maneuvers are the only ones permitted, they are optimal
for these states. We note this fact on the solution grid in
Figure 4.24. There we have noted from the way the flight test
data is presented that quantization increments of $\Delta h = 15,000$ ft.
in altitude and $\Delta v = 0.5$ M in velocity are appropriate. For each
grid point at h = 75,000 ft., we have indicated that the optimal
strategy is to accelerate by an arrow pointing to the right, and for
each grid point at v = 3.0 M we have indicated that the optimal
strategy is to climb by an arrow pointing upward. The minimum-time-
to-climb for each of these grid points is found by simply adding

the appropriate times from Tables 4.14 and 4.15 corresponding to
the maneuvers required to reach h = 75,000 ft., v = 3.0 M; these
minimum time values are shown beside the grid points in Figure 4.24.

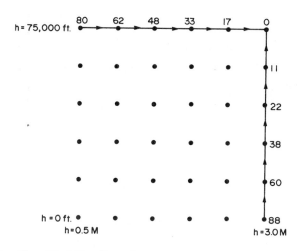

Figure 4.24   Initial Calculation Procedure

Next, we consider the calculation at h = 60,000 ft., v = 2.5 M.
Note that we can either climb to h = 75,000 ft., v = 2.5 M or accelerate to h = 60,000 ft., v = 3.0M.   If we define I(h,v) to be
the minimum time-to-climb from altitude h, velocity v, to altitude
75,000 ft., velocity 3.0 M, we can write down immediately the
dynamic programming recursive equation as

$$I(h,v) = \min \{\Delta t_c(h,v) + I(h + \Delta h, v),$$

$$\Delta t_a(h,v) + I(h,v + \Delta v)\}$$

where Δh = 15,000 ft., Δv = 0.5 M.   For h = 60,000 ft., v = 2.5 M,
this equation is

$$I(60,000, 2.5) = \min \{12 + 17, 19 + 11\}$$
$$= 29$$

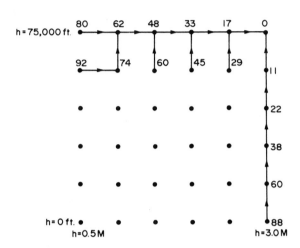

Figure 4.25   Additional Calculations at h = 60,000 ft

The corresponding optimal maneuver is to climb. This result is entered on Figure 4.25.

Continuing, we observe that we can now complete the calculation at h = 60,000 ft., v = 2.0 M. A climb maneuver takes us to the state h = 75,000 ft., v = 2.0 M, a state for which we obtained the minimum-time-to-climb in Figure 4.24 as 33 seconds. An accelerate maneuver takes us to state h = 60,000 ft., v = 2.5 M, the state for which we have finished finding the minimum-time-to-climb as 29 seconds. Using Tables 4.14 and 4.15 we find immediately that

$$I(6,0000, 2.0) = min \quad \{12 + 33, 18 + 29\}$$
$$= 45$$

and that the optimal maneuver is again to climb. This result is also shown in Figure 4.25.

We see that by working from v = 3.0 M in the order of decreasing velocity at altitude h = 60,000, we can obtain the solution for all grid points at this altitude. All these solutions are shown in Figure 4.25.

This procedure can be continued in a straightforward manner,
one row of constant altitude at a time, until the complete solution
has been obtained. This solution is shown in Figure 4.26. From
the figure we see that the minimum-time-to-climb from h=0, v=0.5M
is 170 seconds and that the optimum sequence of maneuvers is to climb
to h = 15,000 ft., v = 0.5 M; accelerate to h = 15,000 ft., v = 1.0 M;
climb to h = 30,000 ft., v = 1.0 M; accelerate to h = 30,000 ft.,
v = 1.5 M; climb three increments in altitude to 75,000 ft.,
v = 1.5 M; and accelerate three increments in velocity to
h = 75,000 ft., v = 3.0 M.

(b) We can use fundamentally the same procedure if the zoom
maneuver is also allowed. For all states at v = 3.0 M, the only
allowed maneuver is to climb. However, for states at h = 75,000 ft.,
v ≠ 3.0 M., there are two allowed maneuvers - accelerate and zoom.
Thus, we can only obtain minimum-time-to-climb for the former
states and not the latter. Our starting condition is thus as
shown in Figure 4.27.

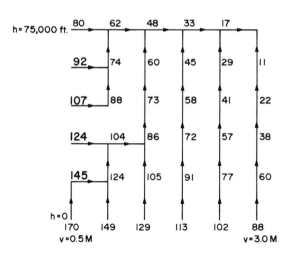

Figure 4.26   Complete Solution Grid

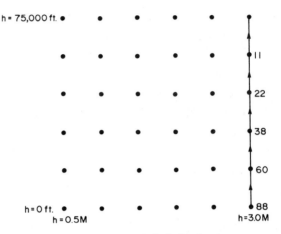

Figure 4.27   Initial Calculations

To perform our calculations with the zoom maneuver included,
we again define $I(h,v)$ as the minimum-time-to-climb from
altitude h, velocity v to altitude 75,000 ft., velocity 3.0 M.
The recursive equation is modified to read

$$I(h,v) = \min \{\Delta t_c(h,v) + I(h + \Delta h, v), \Delta t_a(h,v) + I(h,v + \Delta v),$$

$$\Delta t_z(h,v) + I(h - \Delta h, v + \Delta v)\}$$

where again $\Delta h = 15,000$ ft., $\Delta v = 0.5$ M.

At $h = 75,000$ ft., $v = 2.5$ M, we see that there are only two
possible maneuvers - accelerate and zoom. The recursive equation
then reduces to

$$I(75,000, 2.5) = \min \{17 + 0, 4.5 + 11\}$$
$$= 15.5$$

The optimal maneuver for this state is to zoom. This solution is
shown in Figure 4.28.

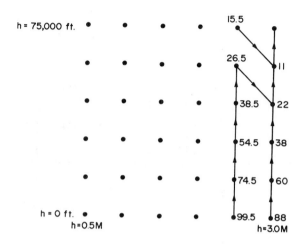

Figure 4.28   Additional Calculations at v = 2.5 M

We next see that it is now possible to compute minimum-time-to climb at h = 60,000 ft., v = 2.5 M. The next state for the maneuver climb is h = 75,000, v = 2.5, the state at which we just completed our calculations. The next states for accelerate and zoom are h = 60,000 ft., v = 3.0 M and h = 45,000 ft., v = 3.0 M respectively. Solutions for both of these states are shown in Figure 4.27. The calculation at this state yields

$$I(60,000, 2.5) = \min \{12 + 15.5, \ 19 + 11, \ 4.5 + 22\}$$
$$= 26.5$$

The optimal maneuver for this state is again to zoom. This solution is shown in Figure 4.28. We can continue to apply this procedure to the states in the column v = 2.5 M, working downward in the direction of decreasing altitude. The solution after completing these calculations is shown in Figure 4.28. This solution can be continued in a straightforward manner, one column of constant velocity at a time, until the complete solution has been obtained. This solution is shown in Figure 4.29. From this figure we see that the minimum-time-to-climb using the additional zoom maneuver

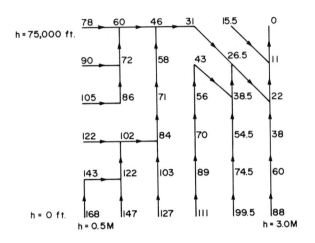

Figure 4.29   Complete Solution Grid

is 168 seconds, 2 seconds less than the minimum-time-to-climb
without it. The optimal sequence of maneuvers is to climb to h =
15,000 ft., v = 0.5 M; accelerate to h = 15,000 ft., v = 1.0 M;
climb to h = 30,000 ft., v = 1.0 M; accelerate to 30,000 ft.,
v = 1.5 M; climb three increments to h = 75,000 ft., v = 2.0 M;
zoom two increments to h = 45,000 ft., v = 3.0 M; and climb two
increments to h = 75,000 ft., v = 3.0 M.

4.5 A harried businessman has just arrived at the airport in a
large metropolitan area. It is the rush hour, and he is running
late for a very important meeting on the other side of town. Fortunately, the girl at the rental car agency is very familiar with
the traffic patterns at this hour on freeways, boulevards, and
back streets. She has constructed the chart in Figure 4.30, which
shows a network of possible routes to the meeting and the driving
time along each link in this network. All streets are one-way
streets, so that he must always move toward the meeting (the right
of one page) and not double back. Use dynamic programming to help
the businessman find the minimum time route to his meeting. The
solution is shown on the grid in Figure 4.31.

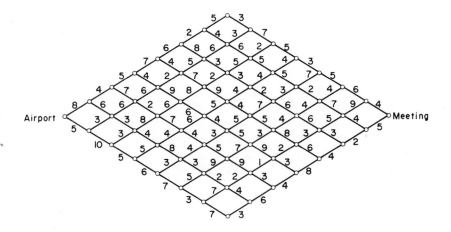

Figure 4.30   Network for Businessman's Minimum
                Time Route Problem

4.6   Consider the hydro-thermal generation scheduling problem shown
in Figure 4.32a. The amount of water in the hydro reservoir at hour
k is denoted as x(k). The power output during hour k from the gene-
rator at the reservior is equal to u(k), the outflow from the
reservoir. The inflow during hour k is N(k), where

        N(k) = 1
        k = 0,1,2,3,4

The power output during hour k from the thermal plant is y(k). As
shown in Figure 4.32b, the cost of producing the output y(k) is

        $C(y(k)) = [y(k)]^2$

The total power demand at hour k, D(k), must be met by the sum of
the outputs from the hydro plant and the thermal plant. The demand
over hours k = 0,1,2,3,4 is shown in Table 4.17. The amount of
water in the reservoir is constrained by $0 \leq x \leq 2$, and it is
constrained to be 1 at hour 0 and hour 5. The release from the
reservoir is constrained by $0 \leq u \leq 2$. Using quantization increments of $\Delta x = 1$ and $\Delta u = 1$, apply the forward dynamic programming
computational procedure to obtain the optimal operation of the
reservoir over hours k = 0,1,2,3,4.

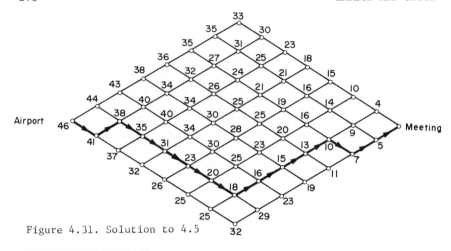

Figure 4.31. Solution to 4.5

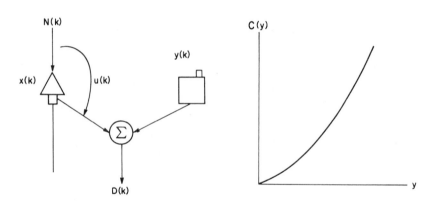

Figure 4.32a. Hydro Thermal System   4.32b. Cost Curve for Thermal
                                                    Plant

Table 4.17   Power Demand for Total System

k	D(k)
0	2
1	3
2	2
3	1
4	2

From our previous analysis of hydro thermal scheduling problems
in Chapter 2, we recall that x(k), the water content of the reservoir,
is the state variable; u(k), the release from the reservoir, is the
decision variable; and k, the hour, is the stage variable. We then
see that the water balance equation for the reservoir is

$$x(k+1) = x(k) + 1 - u(k)$$
$$k = 0,1,2,3,4$$

where we have noted that the inflows N(k), k = 0,1,2,3,4, are all
equal to 1. This equation becomes the system equation for the
problem. The constraints on the state and decision variables are
easily obtained as

$$0 \le x \le 2, \ k = 1,2,3,4$$
$$x(0) = 1 \ ,$$
$$x(5) = 1 \ ,$$
$$0 \le u \le 2.$$

The performance criterion, which is the reduction in thermal cost
obtained from using the hydro, must be expressed in terms of the
state and decision variables. We can use the known total system
demand and the thermal cost curve to obtain the required relations.
First we see that

$$D(k) = y(k) + u(k)$$

Therefore,

$$y(k) = D(k) - u(k)$$

The thermal cost is given as

$$C(y(k)) = [y(k)]^2 = [D(k) - u(k)]^2.$$

Therefore, the desired quantity to be optimized is

278   LARSON AND CASTI

$$J = \sum_{k=0}^{4} [D(k) - u(k)]^2$$

where D(k) is given in Table 4.17.

If we use the prescribed quantization increments of $\Delta x=1$ and $\Delta u=1$ and the constraints, we see that the set of admissible states at stages k = 0,1,2,3,4,5 is given by

$$X(0) = X(5) = \{1\}$$
$$X(k) = \{0,1,2\} , k = 1,2,3,4,$$

while the set of admissible decisions is given by

$$U = \{-1, 0, 1\}$$

at all stages.

To apply the forward dynamic programming computational procedure, we set up the grid shown in Figure 4.33. The minimum cost function I'(x,0) has the value 0 at the prescribed initial state x=1, while all other initial states are inadmissible, as shown by the x's in the figure.

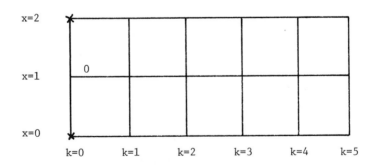

Figure 4.33   Starting Condition for Computational
              Procedure

We next apply the normal computational algorithm noting that the inverse function $g^{-1}(x(k+1), u(k), k)$ is given by

$$g^{-1}(x(k+1), u(k), k) = x(k+1) -1 + u(k)$$

Since this inverse is straightforward to compute, we can store the optimum previous decision $\hat{u}'(x,k)$ directly rather than the optimum previous state. We begin the computational procedure at stage 0. We apply all possible decisions for the single admissible state $x(0) = 1$. The system equation is $x(1) = 2-u$. We calculate the corresponding single stage cost as

$$L(1,u,0) = [2-u]^2$$

Since there is only one admissible state at stage 0, the tentative minimum costs are immediately the actual minimum costs. The results are shown in Figure 4.34, where the number above and to the right of the grid point is the minimum cost and the number below and to the right is the optimum previous decision.

At stage 2 we have to perform a comparison. For $x(1)=2$, we see that the system equation is

$$x(2) = 3-u$$

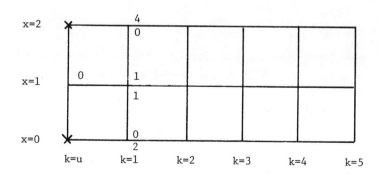

Figure 4.34   Results of Calculations at Stage 1

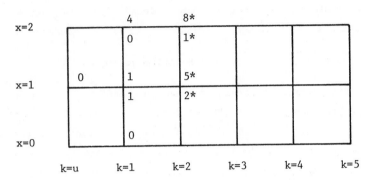

Figure 4.35   Results at k=2 for state x(1) = 2

and that the single stage cost is

$$L(2,u,1) = [3-u]^2$$

We obtain the tentative minimum costs and previous decisions shown in Figure 4.35 above, where the asterisks on the quantities indicate that they are tentative. For x(1) = 1, the system equation is

$$x(2) = 2-u$$

and the single-stage cost function is

$$L(1,u,k) = [3 - u]^2$$

The results of applying all decisions for this state are shown in Figure 4.36. For x(1) = 0, the system equation is

$$x(2) = 1-u$$

and the single-stage cost function is

$$L(0,u,1) = [3-u]^2$$

The results of applying the decisions for this state are shown in Figure 4.37. Since this is the last admissible state at stage k=1, all minimum costs and previous decisions are final, not tentative.

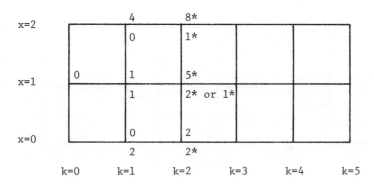

Figure 4.36   Results at k=2 for States x(1)=2
and x(1)=1

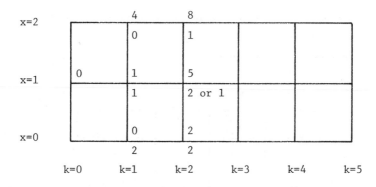

Figure 4.37   Results at k=2 for States x(1)=2,
x(1)=1, and x(1)=0

The calculations continue in this manner. The only change at each
stage is the single-stage cost function. We see that

$$L\ (x,u,2) = [2-u]^2$$
$$L\ (x,u,3) = [1-u]^2$$
$$L\ (x,u,4) = [2-u]^2$$

The results of the calculations are shown in Figure 4.38

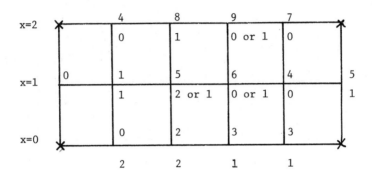

Figure 4.38   Complete Results

The solution can be recovered backwards in k using the previous decision from the figure.   The resulting state trajectory and decision sequence is shown in Table 4.18.

Table 4.18   Optimal Solution

k	x	u	L(x,u,k)
0	1	1	1
1	1	2	1
2	0	1	1
3	0	0	1
4	1	1	1
5	1	-	0
Total			5

The resulting hydro and thermal generation schedules are shown in Table 4.19.   This result agrees with a principle from electric utility operation that, if the thermal cost function is convex, then the hydro generation should be adjusted so as to hold the thermal generation as nearly constant as possible.   In this case the inflows and reservoir constraints are such that it can be held exactly constant.

Table 4.19   Hydro Generation, Thermal Generation,
              and Demand Corresponding to the Optimal
              Schedule

k	Hydro Generation	Thermal Generation, y	Demand, D
0	1	1	2
1	2	1	3
2	1	1	2
3	0	1	1
4	1	1	2

4.7   A student at a major Western university is about to take an
important examination.   The exam covers questions from four categories,
and he is given a grade from 1 to 10 in each category.   His expected
score in each category is as follows:

           Systems Optimization   7
           Systems Modeling   4
           Economics   3
           Applications   6

He is allowed to weight the score in each category by an integer
from 1 to 5 inclusive.   The sum of the weights must equal 10.
(a)   Develop a forward dynamic programming procedure to find the set
of weights that maximizes his expected score.
(b)   Apply the procedure to find the optimum weights for this case.

We can modify our usual resource allocation formulation.   Let
task 1 be Systems Optimization, task 2 be Systems Modeling, task 3
be Economics, and task 4 be Applications.   If u is the weight assigned
to any task, then we see that

$$R(u,1) = 7u$$
$$R(u,2) = 4u$$
$$R(u,3) = 3u$$
$$R(u,4) = 6u$$

Since $1 \leq u \leq 5$ and u must be an integer, we have $U = \{1,2,3,4,5\}$. By implication, $X = \{1,2,3,4,5,6,7,8,9,10\}$. We then solve the equation

$$I(x,0) = 0 \quad , x = 10$$
$$= -\infty , x \neq 10$$

$$I(x,k) = \underset{u}{\text{Max}} \{R(u,k) + I(x+u, k-1)\}$$

$$0 \leq x \leq 10$$

$$k = 1,2,3 .$$

The solution grid is as shown.

x=10	x	x	x	x
9	7,1	x	x	x
8	14,2	11,1	x	x
7	21,3	18,1	14,1	x
6	28,4	25,1	21,1	20,1
5	35,5	32,1	28,1	27,1
4	x	39,1	35,1	34,1
3	x	43,2	42,1	41,1
2	x	47,3	46,1	48,1
1	x	51,4	50,1	54,2
0	x	55,5	54,1	60,3
	k=1	k=2	k=3	k=4

The optimal allocation is 5 to task 1, 1 to task 2, 1 to task 3 and 3 to task 4, with an expected score of 60.

SUPPLEMENTARY PROBLEMS

4.8   Consider further the aircraft minimum-time-to-climb problem
of Problem 4.4.

(a)   In addition to the climb, accelerate, and zoom maneuvers
of Problem 4.4, the aircraft is also allowed to "zap", where a zap
is defined to be a climb of 15,000 ft. during which the aircraft
increases its speed by 0.5M. Assume that the time to complete a
zap is given by Δt_b in Table 4.20. Extend the procedure of
Problem 4.4 to solve the minimum-time-to-climb problem while
working in a two-dimensional grid such as that portrayed in Figures
4.26 and 4.29. Use this procedure to re-compute a new minimum time
and corresponding sequence of maneuvers to reach h=75,000 ft. and
v=3.0M from h=0, v=0.5M.

(b)   In addition to the climb, accelerate, zoom, and zap maneuvers
previously defined, the aircraft is allowed to "zing", where zing
is defined to be a climb of 15,000 ft. during which the aircraft
decreases its speed by 0.5M. Assume that the time given to complete
a zing is given by Δt_d in Table 4.20. Extend the procedure of
Problem 4.4 to solve the minimum-time-to-climb problem for this
case while working in a two-dimensional grid such as that portrayed
in Figures 4.26 and 4.29. Use this procedure to re-compute a
new minimum time and corresponding sequence of maneuvers to reach
h-75,000 ft., v=3.0M from h=0, v=0.5M.

Table 4.20 $\Delta t_b(h,v)$ = Time to Climb 15,000 ft. and
Accelerate 0.5M from Altitude h and
Velocity v

h in ft.

60,000	32	30	28	27	26
45,000	34	32	30	29	28
30,000	40	48	36	34	32
15,000	54	50	46	43	40

v in Mach   0.5   1.0   1.5   2.0   2.5
no.

286

LARSON AND CASTI

Table 4.21 $\Delta t_d(h,v)$ = Time to Climb 15,000 ft. and Decelerate to 0.5M. from Altitude h and Velocity v

h in ft.

60,000	6	5	4.5	4	4
45,000	6	5	4.5	4	4
30,000	7	6	5	4.5	4
15,000	8	7	6	5	4.5

v in Mach. 1.0   1.5   2.0   2.5   3.0
no.

4.9 Suppose we have N distinct locations that we must visit and a matrix $T = (t_{ij})$ giving us the time required to travel from location i to location j, with $t_{ii} = 0$. Starting at location 1, we may use any of the other locations as intermediate stops on our way to N. Let $I(i)$ denote the minimal time required to go from i to N, $i=1,2,\ldots,N-1$.

(a) Show that

$$I(i) = \min_{j \neq i} [t_{ij} + I(j)], \quad i=1,2,\ldots,N-1$$

$$I(N) = 0.$$

(b) Show the above equation has a solution $\{I(i)\}$ which is unique up to an additive constant and that any one of these solutions will yield the optimal routing policy.

4.10 Captain Kork is flying the Starship Expertise from Andromeda to Orion. The normal flight mode of the Expertise is characterized by the system equations

$$x(k+1) = x(k) + u(k),$$

and constraints

$$0 \leq x \leq 2,$$
$$-1 \leq u \leq 1,$$

where x(k) is the stellar state at stage k and u(k) is the stellar command at stage k. The location of Andromeda is specified by initial stellar state

$$x(0) = 2$$

while the location of Orion is specified by terminal stellar state

$$x(6) = 2.$$

As in earthly applications, it is convenient to quantize x and u in unit increments ($\Delta x = 1$, $\Delta u = 1$), so that

$$X = \{0,1,2\}$$
$$U = \{-1,0,1\}$$

The cost of flight in this mode is given by

$$J = \sum_{k=0}^{5} [x^2(k) + u^2(k)] .$$

The Expertise can also fly in the space warp mode; in this mode the Expertise maintains constant stellar state but moves either 3,4,5 or 6 stages ahead. The cost of a space warp at stage x, stage k that lasts n stages results in a stellar state x at stage (k+n) and a cost of

$$L_W = \frac{1}{2} + \frac{1}{2} n x^2 .$$

Find the least-cost flight path in stellar state space from Andromeda to Orion, using both conventional stellar commands and space warp.

4.11 Consider the problem with cost function

$$J = \sum_{k=1}^{N-1} L(x(k),u(k),k)$$

system equation

$$x(k+1) = g(x(k), u(k), k), \quad k=0, p, \ldots, K-1,$$

constraints $x \in X \subset R^n$ and $u \in U \subset R^m$, fixed initial state $x(0) = c$ and fixed terminal state $x(K) = d$. Prove that the complete forward dynamic programming approach and the backward dynamic programming procedure both determine the same optimum decision sequence and optimal trajectory. (HINT: Use proof by contradiction as in the section on the principle of optimality in Chapter 2).

4.12 Recall the problem of Example 2.4 (Example within the Chapter, not solved problem at the end of the chapter.), which was solved explicitly using the backward dynamic programming approach. The optimal decision sequence and optimal trajectory from state $x(0) = 1$ were obtained in Example 2.7 (Example within the Chapter, not solved problem at the end of the chapter.). Use both forward dynamic programming approaches to solve this problem for fixed initial state $x(0) = 1$. How do the answers compare with the backward dynamic programming solution? Explain.

4.13 Refer to the aerospace vehicle in Solved Problem 2.6. Assume that the vehicle described there is part of an antiballistic missile system. The vehicle has an initial state consisting of a known ground location with zero velocity. Its mission is to intercept a ballistic missile. At a prescribed initial time a ballistic missile trajectory is made known to the vehicle; this trajectory specifies the missile's position and velocity in the coordinate system of Problem 2.4 as a function of time. The vehicle must then determine a launch time, a launch trajectory, and a warhead detonation time to respond to this trajectory. The criterion function for this response is a function of the quality of intercept at detonation time; this criterion applies relative weightings to miss distance (negative weighting), relative angle of velocity vectors (negative weighting), and altitude (positive weighting) at the detonation time.

(a) Develop a complete formulation for this problem using the

notation of Solved Problem 2.6. Let this ballistic missile trajectory

be represented by $[y_1(k), y_2(k), y_3(k), y_4(k)]$, $k=0,\ldots,K$, where

stage 0 represents the prescribed initial time and stage K represents

the time of impact on the ground.

(b) Develop a complete methodology for solving the problem using

the results of a forward dynamic programming calculation. Your

methodology should be such that all vehicle trajectory optimization

can be done <u>prior</u> to the time when the launch decision must be made.

(Hint - see Ref. L-13).

4.14 Recall the chemical process of Solved Problem 2.7. Consider

the modified chemical process in Figure 4.39. This process is identi-

cal to that of Problem 2.6 from Reactor 4 onward. However, the two

inputs to Reactor 4 come from separate parallel processes.

A unit amount of Chemical B at concentration C_{BO} is processed

in Reactor 3 along with a catalyst at concentration D_3 at tempera-

ture T_3. The resulting chemical at concentration C_{B3} is a function

of input concentration C_{BO}, catalyst concentration D_3, and tempera-

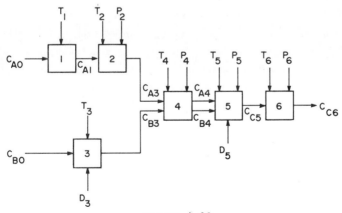

Figure 4.39

ture T_3 according to $C_{B3} = f_3(C_{BO},D_3,T_3)$. The cost of the catalyst
is a function of its concentration $\phi_{D3}(D_3)$, while the cost of
reactor operation is $\phi_{C3}(C_{BO},D_3,T_3)$.

At the same time a unit amount of Chemical A at concentration
C_{AO} is processed in Reactor 1 at a temperature T_1. The concentration of this chemical becomes C_{A1} according to the function C_{A1}
$= f_1(C_{AO},T_1)$. The cost of operating this reactor is $\phi_1(C_{AO},T_1)$.
The resulting chemical is then fed to Reactor 2 at a temperature
T_2 and pressure P_2, where its concentration is modified to C_{A3}
according to the relation $C_{A3} = f_2(C_{A1},T_2,P_2)$. The cost of operating
this reactor is $\phi_2(C_{A1},T_2,P_2)$. The output of this reactor, chemical
A at concentration C_{A3}, is the other input to Reactor 4.

Use a combination of backward and forward dynamic programming
to obtain an efficient computational procedure for this problem.
Use the approach in Solved Problem 2.7 to calculate $I(x_1,x_2,3)$,
the maximum profit for going to the end of the process from the input
to Reactor 4, given that at this point the concentration of chemical
A is x_1 and the concentration of chemical B is x_2. Next, use foward
dynamic programming to find $I'(x_2,3)$, the maximum profit (least
cost) for producing concentration x_2 of chemical B at the input to
reactor 3 starting at the process input. Similarly, use forward
dynamic programming to find $I'(x_1,3)$, the maximum profit (least cost)
for producing concentration x_1 of chemical A at the input to Reactor
3 starting at the process input. Finally, calculate the optimum
values of x_1 and x_2 at this point by maximizing

$$\{I'(x_1,3) + I'(x_2,3) + I(x_1,x_2,3)\}.$$

4.15 Consider the following airline schedule problem. An aircraft can take any of the flights shown in Table 4.22. The profit and duration of each flight is as shown in the Table. The aircraft cannot fly between cities, except on one of these flights. Find the optimum sequence of flights over a 24-hour period if

(a) The aircraft can start at time 0 at any airport and end at time 24 at any airport.

(b) The aircraft can start at time 0 at any airport but must be in ORD at time 24 for maintenance.

Hint - if you use forward dynamic programming, part (b) will be very easy to solve after you have solved part (a).

Table 4.22   Scheduled Flights

Flt. No.	Origin/ Destination	Departure Time	Arrival Time	Profit
1	SFO/ORD	0	4	150
2	JFK/ORD	2	4	100
3	ORD/JFK	5	7	200
4	ORD/SFO	5	9	300
5	JFK/SFO	8	14	500
6	ORD/JFK	9	11	250
7	SFO/JFK	9	15	450
8	JFK/ORD	12	14	200
9	ORD/JFK	14	16	200
10	SFO/JFK	15	21	400
11	JFK/SFO	17	23	500
12	JFK/ORD	18	20	300

REFERENCES

Problems with an Implicit Stage Variable: [B-44], [B-77], [B-89],
                                            [B-93], [B-104], [D-14]

Infinite-Stage Processes: [L-17], [V-2], [D-2], [B-60]

Forward Dynamic Programming: [L-13], [L-14]

REFERENCES

[A-1]   J. Aczel, Lectures on Functional Equations and Their Appli-
         cations, Academic Press, New York 1966.

[A-2]   S. Altshuler, "Variational Principles for the Wave Equation
         Function Theory," Phys. Rev., 109, 1958, pp. 1830-1836.

[A-3]   E. Angel, "Dynamic Programming and Partial Differential
         Equations," USCEE-265, Univ. of Southern California,
         Los Angeles, March 1968.

[A-4]   E. Angel, "A Building Block Technique for Elliptic Boundary-
         value Problems over Irregular Regions," USCEE-285, Univ. of
         Southern California, Los Angeles, June 1968.

[A-5]   E. Angel, "Discrete Invariant Imbedding and Elliptic Boundary-
         value Problems over Irregular Regions," USCEE-279, Univ. of
         Southern California, Los Angeles, June 1968.

[A-6]   E. Angel, "Noniterative Solutions of Nonlinear Elliptic
         Equations," USCEE-325, Univ. of Southern California, Los
         Angeles, October 1968.

[A-7]   E. Angel, "Invariant Imbedding and Three-dimensional Poten-
         tial Problems," USCEE-325, Univ. of Southern California,
         Los Angeles, January 1969.

[A-8]   E. Angel, "Inverse Boundary-value Problems: Elliptic Equations,"
         USCEE-343, Univ. of Southern California, Los Angeles,
         April 1969.

[A-9]   M. Aoki, "On the Approximation of Trajectories and its Appli-
         cation to Control Theory Optimization Problems," JMAA, 9,
         1964, pp. 23-41.

[A-10]   M. Aoki, "On Optimal and Suboptimal Policies in Control
         Systems," Chapt. 1 in Advances in Control Systems, Vol. I;
         Theory and Applications, (C. T. Leondes, ed.), Academic
         Press, New York 1964.

[A-11]   M. Aoki, "Optimal Bayesian and Min-Max Controls of a Class
         of Stochastic and Adaptive Dynamic Systems," Preprints,
         pp. II-21-II-31, paper presented at IFAC Tokyo Symposium
         on Systems Engineering for Control System Design, Tokyo,
         Japan (September 1965).

293

[A-12] M. Aoki, "Control of Large Dynamic Systems by Aggregation,"
        IEEE Trans. Autom. Control , June 1968.

[A-13] R. Aris, G. L. Nemhauser, and D. J. Wilde, "Optimization of
        Multistage Cyclic and Branching Systems by Serial Procedures,"
        A.I.Ch.E.J. 10, 1964, p. 913.

[A-14] R. Aris, Discrete Dynamic Programming, Blaisdell, New York,
        1964.

[A-15] R. F. Arnold, D. L. Richards, "Monotone Reduction Algorithms,"
        in Recent Mathematical Advances in Operations Research, No.
        129, Univ. of Michigan, Ann Arbor, Summer 1964.

[A-16] G. Aronsson, "Minimization Problems for the Functional sup
        $F(x,f(x),f'(x))$," Ark. Mat. 6, 1966, pp. 409-431.

[A-17] G. Aronsson, "Minimization Problems for the Functional sup
        $F(x,f(x),f'(x))$-II," Ark. Mat.6, 1966, pp. 409-431.

[A-18] K. D. Arrow, T. E. Harris and J. Marshak, "Optimal Inventory
        Policy," Econometrica, vol. 19, 1951, pp. 250-272.

[A-19] K. J. Arrow, S. Karlin and H. Scarf, Studies in the Mathema-
        tical Theory of Inventory and Production, Stanford University
        Press, Stanford, California, 1958.

[A-20] K. J. Arrow, "Applications of Control Theory to Economic
        Growth," AMS Lec. Appl. Math, 12, pp. 85-119, 1968.

[A-21] M. Ash, Optimal Shutdown Control of Nuclear Reactors, Aca-
        demic Press, New York, 1966.

[A-22] M. Athans and P. L. Falb, Optimal Control, McGraw-Hill, 1966.

[A-23] M. Avriel and D. J. Wilde, "Golden Block Search for the
        Maximum of a Unimodal Functions," Manag. Sci., 14, 1968,
        pp. 307-319.

[A-24] S. Azen, Higher Order Approximation to the Computational
        Solution of Partial Differential Equations, The RAND
        Corp., RM-3917, 1964.

[A-25] S. P. Azen, Successive Approximations by Quadratic Fitting
        as Applied to Optimization Problems, The RAND Corp.,
        RM-3301-PR, 1966

[B-1]   I. Babushka and S. L. Sobolev, "Optimization of Numerical
        Processes" (in Russian), Appl. Math., 10, 1965, pp. 96-129.

[B-2]   P. B. Bailey and L. Shampine, Nonlinear Two-point Boundary-
        value Problems, Academic Press, New York, 1969.

[B-3]   J. F. Baldwin and J. S. Sims-Williams, "An On-Line Control
        Scheme Using A Successive Approximations in Policy Space
        Approach," J. Math. Anal. Appl., 22, 1968, pp. 523-536.

[B-4]   H. T. Banks, "Variational Problems Involving Functional
        Differential Equations," J. SIAM Contr., 7, 1969, pp. 1-17.

[B-5]   E. F. Beckenbach and R. Bellman, Inequalities, Springer,
        Berlin, 1961.

[B-6]   R. E. Beckwith, Analytic and Computational Aspects of Dynamic
        Programming Processes of High Dimension, JPL, California
        Institute of Technology, Pasadena, 1959, pp. 1-125.

[B-7]   R. Bellman, I. Glicksberg and O. Gross, "On the Optimal
        Inventory Equation", Management Science, Vol. 2, 1955,
        pp. 83-104.

[B-8]   R. Bellman, "Some Functional Equations in the Theory of
        Dynamic Programming - I: Functions of Points and Point
        Transformations," Trans. Amer. Math. Soc., 80, 1955, pp. 51-71.

[B-9]   R. Bellman, I. Glicksberg and O. Gross, "On the Bang-Bang
        Control Problem," Q. Appl. Math., Vol. 14, 1956, pp. 11-18.

[B-10] R. Bellman, "A Variational Problem with Constraints in
        Dynamic Programming," J. Soc. Indus. Appl. Math., 4,
        1956, pp. 48-61.

[B-11] R. Bellman, "Mathematical Aspects of Scheduling Theory,"
        J. Soc. Ind. Appl. Math., 4, 168-205 (1956).

[B-12] R. Bellman, "Dynamic Programming and Its Application to
        Variational Problems in Mathematical Economics," Proc.
        Symp. Calculus of Variations and Appli., April 1956,
        pp. 115-138.

[B-13] R. Bellman, I. Glicksberg and O. Gross, "Some Nonclassical
        Problems in the Calculus of Variations," Proc. Amer. Math.
        Soc., 7, 1956, pp. 87-94.

[B-14] R. Bellman, "On the Computational Solution of Linear Program-
        ming Problems Involving Almost-Block Diagnoal Matrices,"
        Management Sci., 1957, pp. 403, 406.

[B-15] R. Bellman, Dynamic Programming, Princeton University Press,
        N. J. 1957.

[B-16] R. Bellman, "Terminal Control, Time Lags, and Dynamic
        Programming," Proc. Nat. Acad. Sci. USA, 43, 1957,
        pp. 927-930.

[B-17] R. Bellman, "Functional Equations in the Theory of Dynamic
        Programming - VII: A Partial Differential Equation for the
        Fredholm Resolvent," Proc. Amer. Math. Soc., 8, 1957,
        pp. 435-440.

[B-18] R. Bellman, "Functional Equations in the Theory of Dynamic
        Programming-VI: A Direct Convergence Proof," Ann. Math.,
        65, 1957, pp. 215-223.

[B-19] R. Bellman, "On Monotone Convergence to Solutions of
        u' = g(u,t)," Proc. Amer. Math Soc., 8, 1957, pp. 1007-1009.

[B-20] R. Bellman, "On a Class of Variational Problems," Quart.
        Appl. Math., 14, 1957, pp. 353-39.

[B-21] R. Bellman and H. Osborn, "Dynamic Programming and the
        Variation of Green's Functions," J. Math. Mech., 7, 1958,
        pp. 81-86.

[B-22] R. Bellman, I. Glicksberg, and O. Gross, Some Aspects of the
        Mathematical Theory of Control Processes, The RAND Corp.
        R-313, 1958.

[B-23] R. Bellman, "On a Routing Problem," Quart Appl. Math., 16,
        1958, pp. 87-90.

[B-24] R. Bellman, I. Cherry, and G. M. Wing, "A Note on the Numer-
        ical Integration of a Class of Nonlinear Hyperbolic Equations,"
        Quart. Appl. Math., 16, 1958, pp. 181-183.

[B-25] R. Bellman, "Functional Equations in the Theory of Dynamic
        Programming-XI: Limit Theorems," Rend. Circ. Mat. Palermo,
        8, 1959, pp. 1-3.

[B-26] R. Bellman, M. Ash, and R. Kalaba, "On Control of Reactor
        Shutdown Involving Minimal Xenon Poisoning," Nuclear Sci.
        Eng., 6, 1959, pp. 152-156.

[B-27] R. Bellman, J. Jacquez and R. Kalaba, "Some Mathematical
        Aspects of Chemotherapy - I: One-Organ Models," Bull. Math.
        Biophys., 22, 1960, pp. 181-198.

[B-28] R. Bellman, J. Jacquez and R. Kalaba, "The Distribution of a
        Drug in the Body," Bull. Math. Biophys., 22, 1960, pp. 309-322

[B-29] R. Bellman and P. Brock, "On the Concepts of a Problem and
        Problem-Solving," Amer. Math. Monthly, 67, 1960, pp 119-134.

[B-30]   R. Bellman, "Combinatorial Processes and Dynamic Programming,"
         Proc. Ninth Symp. Appl. Math., American Mathematical Society,
         1960.

[B-31]   R. Bellman and R. Kalaba, "On k-th Best Policies," J. Soc.
         Indus. Appl. Math., 8, 1960, pp. 582-588.

[B-32]   R. Bellman, Adaptive Control Processes, Princeton University
         Press, Princeton, N. J., 1961.

[B-33]   R. Bellman, "Successive Approximations and Computer Storage
         Problems in Ordinary Differential Equations," Comm. Assoc.
         Comput. Machinery, 4, 1961, pp. 222-223.

[B-34]   R. Bellman, "On the Reduction of Dimensionality for Classes
         of Dynamic Programming Processes," J. Math. Anal. Appl., 3,
         1961, pp. 358-360.

[B-35]   R. Bellman and R. Kalaba, "Reduction of Dimensionality,
         Dynamic Programming and Control Processes," J. Basic. Eng.,

[B-36]   R. Bellman and W. Karush, "On a New Functional Transform in
         Analysis: The Maximum Transform," Bull. Amer. Math. Soc., 67,
         1961, pp. 501-503.

[B-37]   R. Bellman, "On the Computational Solution of Differential-
         difference Equations," J. Math. Anal. Appl., 2, 1961, pp.
         108-110.

[B-38]   R. Bellman, "On the Reduction of Dimensionality for Classes
         of Dynamic Programming Processes," J. Math. Anal. Appl., 23,
         1961, pp. 358-360.

[B-39]   R. Bellman, "On the Approximation of Curves by Line Segments
         Using Dynamic Programming," Comm. Assoc. Comput. Machinery,
         4, 1961, pp. 284.

[B-40]   R. Bellman and J. M. Richardson, "A Note on an Inverse
         Problem in Mathematical Physics," Quart. Appl. Math., 19,
         1961, pp. 269-271.

[B-41]   R. Bellman, S. Dreyfus, Applied Dynamic Programming, Prince-
         ton University Press, Princeton, N. Y. 1962.

[B-42]   R. Bellman, ed., Proc. Symp. Appl. Math. in Mathematical
         Problems in the Biological Sciences, Vol. 14, American
         Mathematical Society, 1962.

[B-43]   R. Bellman and R. Kalaba, "Dynamic Programming Applied to
         Control Processes Governed by General Functional Equations,"
         Proc. Nat. Acad. Sci. USA, 48, 1962, pp. 1735-1737.

[B-44] R. Bellman, "Dynamic Programming Treatment of the Traveling
        Salesman Problem," J. Assoc. Comput. Mach.,9, 1962, pp. 61-63.

[B-45] R. Bellman and W. Karush, "On the Maximum Transform and
        Semigroups of Transformations," Bull. Amer. Math. Soc., 68,
        1962, pp. 516-518.

[B-46] R. Bellman, "Quasilinearization and Upper and Lower Bounds
        for Variational Problems," Quart. Appl. Math., 19, 1962,
        pp. 349-350.

[B-47] R. Bellman, R. Kalaba and B. Kotkin, "On a New Approach to
        the Computational Solution of Partial Differential Equations·,"
        Proc. Nat. Acad. Sci. USA, 48, 1962, pp. 1325-1327.

[B-48] R. Bellman, "Some Questions Concerning Difference Approxi-
        mations to Partial Differential Equations," Bull. UMI, 17,
        1962, pp. 118-190.

[B-49] R. Bellman, "From Chemotherapy to Computers to Trajectories,"
        Math. Prob. Biol. Sci., 1962, pp. 225-232.

[B-50] R. Bellman and W. Karush, "On the Maximum Transform,"
        J. Math. Anal. Appl., 6, 1963, pp. 67-74.

[B-51] R. Bellman and W. Karush, "Functional Equations in the Theory
        of Dynamic Programming-XII: An Application of the Maximum
        Transform," J. Math. Anal. Appl.,6, 1963, pp. 155-157.

[B-52] R. Bellman, "A Note on Dynamic Programming and Perturbation
        Theory," Nonlinear Vibration Problems, 1963, pp. 242-244.

[B-53] R. Bellman and R. Kalaba, "Invariant Imbedding and the Inte-
        gration of Hamilton's Equations," Rend. Circ. Mat. Palermo,
        12, 1963, pp. 1-11.

[B-54] R. Bellman and R. Kalaba, "A Note on Hamilton's Equations
        and Invariant Imbedding," Quart. Appl. Math., 21, 1963,
        pp. 166-168.

[B-55] R. Bellman and R. Kalaba, "A Note on Nonlinear Summability
        Techniques in Invariant Imbedding," J. Math. Appl. 6, 1963,
        pp. 465-472.

[B-56] R. Bellman and R. Kalaba, "An Inverse Problem in Dynamic
        Programming and Automatic Control," J. Math. Anal. Appl.,7,
        1963, pp. 322-325.

[B-57] R. Bellman,"Mathematical Model-making as an Adaptive Control
        Process," in Mathematical Optimization Techniques, Univ. of
        California, Press, 1963, pp. 333-339.

[B-58]   R. Bellman and R. Karush, "Dynamic Programming: A Bibliography
         of Theory and Application," The RAND Corp Memo RM-3951-PR,
         1964.

[B-59]   R. Bellman, "Dynamic Programming and Markovian Decision
         Processes with Particular Applications to Baseball and Chess,"
         in Applied Combinatorial Mathematics, Wiley, New York 1964,
         pp. 221-236.

[B-60]   R. Bellman and R. Bucy, "Asymptotic Control Theory," J.
         SIAM Control, 2, 1964, pp. 11-18.

[B-61]   R. Bellman and R. Kalaba (eds.) Mathematical Trends in
         Control Theory, Dover, New York 1964.

[B-62]   R. Bellman, B. Gluss and R. Roth, "On the Identification of
         Systems and the Unscrambling of Data: Some Problems Suggested
         by Neurophysiology," Proc. Nat. Acad. Sci. USA, 52, 1964,
         pp. 1239-1249.

[B-63]   R. Bellman, "Control Theory," Sci. Amer., Sept. 1964,
         pp. 186-200.

[B-64]   R. Bellman and R. Kalaba, "Dynamic Programming, Invariant
         Imbedding and Quasilinearization: Comparisons and Inter-
         connections," in Computing Methods in Optimization Problems,
         ed. by A. Balakrishnan and L. Neustadt, Academic Press, 1965.

[B-65]   R. Bellman and R. Kalaba, Dynamic Programming and Modern
         Control Theory, Academic Press, New York, 1965.

[B-66]   R. Bellman and T. Brown, "Projective Metrics in Dynamic
         Programming," Bull. Amer. Math. Soc., 1965, pp. 773-775.

[B-67]   R. Bellman and R. Kalaba, Quasilinearization and Nonlinear
         Boundary-Value Problems, American Elsevier Publishing Co.,
         Inc., New York, N. Y. 1965.

[B-68]   R. Bellman and R. Kalaba, "On a New Approach to the Numerical
         Solution of a Class of Partial Differential-integral Equations
         of Transport Theory," Proc. Nat. Acad. Sci. USA, 54, 1965,
         pp. 1293-1296.

[B-69]   R. Bellman and R. S. Lehman, "Invariant Imbedding, Particle
         Interaction and Conservation Relations," J. Math. Anal.
         Appl., 10, 1965, pp. 112-122.

[B-70]   R. Bellman and K. L. Cooke, "Existence and Uniqueness Theo-
         rems in Invariant Imbedding-II; Convergence of a New Difference
         Algorithm," J. Math. Anal. Appl., 12, 1965, pp. 247-253.

[B-71] R. Bellman, H. Kagwada and R. Kalaba, "Nonlinear Extrapola-
        tion and Two-point Boundary-value Problems," Comm. A.C.M.,
        1965, pp. 511-512.

[B-72] R. Bellman and K. L. Cooke, "On the Computational Solution
        of a Class of Functional Differential Equations," J. Math.
        Anal. Appl., 12, 1965, pp. 495-500.

[B-73] R. Bellman, "Young's Inequality and the Problem of the Optimum
        Transversal Contous," in Theory of Optimum Aerodynamics Shape
        (A. Miele, ed.) ,Academic Press, New York 1965, pp. 315-317.

[B-74] R. Bellman, "Dynamic Programming, Generalized States and
        Switching Systems," J. Math. Anal. Appl., 12, 1965, pp.
        360-363.

[B-75] R. Bellman, "Segmental Differential Approximation and the
        'Black Box' Problem," J. Math. Anal. Appl., 12, 1965, pp.
        91-104.

[B-76] R. Bellman, "Functional Equations," in Handbook of Mathema-
        tical Psychology, Vol. III, Wiley, New York, 1965.

[B-77] R. Bellman, "An Application of Dynamic Programming to the
        Coloring of Maps," ICC Bull. 4, 1965, pp. 3-6.

[B-78] R. Bellman, J. Buell, and R. Kalaba, "Mathematical Experi-
        mentation in Time-lag Modulation," Comm. Assoc. Comput.
        Machinery, 9, 1966, p. 752.

[B-79] R. Bellman, "On Analogues of Poincare-Lyapunov Theory for
        Multipoint Boundary-value Problems-I," J. Math. Anal. Appl.,
        14, 1966, pp. 522-526.

[B-80] R. Bellman, R. Kalaba, and J. Lockett, Numerical Inversion
        of the Laplace Transform, Elsevier, New York, 1966.

[B-81] R. Bellman and R. Roth, "Segmental Differential Approximation
        and Biological Systems: An Analysis of a Metabolic Process,"
        J. Theoret. Biol., 11, 1966, pp. 168-176.

[B-82] R. Bellman, H. Kagiwada, and R. Kalaba, "Dynamic Programming
        and in Inverse Problem in Neutron Transpo-t Theory," Computing
        2, 1967, pp. 5-16.

[B-83] R. Bellman, "Dynamic Programming: A Reluctant Theory," in
        New Methods of Thought and Procedure, (F. Zwicky and A.
        Wilson, Eds.), Springer, Berlin, 1967.

[B-84] R. Bellman, "On the Validity of Truncation for Infinite
        Systems of Ordinary Differential Equations Associated with
        Nonlinear Partial Differential Equations," Math & Phys. Sci.,
        1, 1967, pp. 95-100.

[B-85] R. Bellman, "Stratification and the Control of Large Systems
        with Application to Chess and Checkers," Inform. Sci., 1,
        1968, pp. 7-21.

[B-86] R. Bellman, "Some Inequalities for theSquare Root of a
        Positive Definite Matrix," Linear Algebra Its Appl., 1,
        1968, pp. 321-324.

[B-87] R. Bellman, Introduction to the Mathematical Theory of
        Control Processes, Vol. I: Linear Equations and Quadratic
        Criteria, Academic Press, New York, 1968.

[B-88] R. Bellman, "A New Type of Approximation Leading to Reduction
        of Dimensionality in Control Processes," JMAA, 27, 1969,
        pp. 454-459.

[B-89] R. Bellman and K. L. Cooke, "The Konigsberg Bridges Problem
        Generalized," J. Math. Anal. Appl., 25, 1969, pp. 1-7.

[B-90] R. Bellman, Dynamic Programming and Problem-Solving,
        TR-69-2, Univ. Southern California, 1969.

[B-91] R. Bellman, Introduction to Matrix Analysis, 2nd ed. McGraw-
        Hill, New York, 1970.

[B-92] R. Bellman, Methods of Nonlinear Analysis, Academic Press,
        New York, 1970.

[B-93] R. Bellman, K. L. Cooke and J. Lockett, Algorithms, Graphs
        and Computers, Academic Press, New York, 1970.

[B-94] R. Bellman, "Functional Equations in the Theory of Dynamic
        Programming-XV: Layered Functionals and Partial Differential
        Equations," J. Math. Anal. Appl., to appear.

[B-95] R. Bellman and R. Roth, "A Technique for the Analysis of a
        Broad Class of Biological Systems," Bionics Symp., pp. 725-737.

[B-96] R. Bellman,"A Note on Asymptotic Control Theory," J. Math.
        Anal. Appl., to appear.

[B-97] R. Bellman, "On Analogues of Poincare-Lyapunov Theory for
        Multipoint Boundary-value Problems-Correction," J. Math.
        Anal. Appl., to appear.

[B-98] R. Bellman and J. Casti, "Differential Quadrature and Long-
        Term Integration," JMAA, 34, 1971, pp. 235-236.

[B-99] R. Bellman, "A Function is a Mapping – Plus a Class of
        Algorithms," Inform. Sci., to appear.

[B-100] R. Bellman, "Dynamic Programming and Inverse Optimal Problems in Mathematical Economics," J. Math. Anal. Appl., to appear.

[B-101] L. D. Berkowitz, "Variational Methods in Problems of Control and Programming," J. Math. Anal. Appl., 3, 1961, pp. 145-169.

[B-102] L. D. Berkowitz, "A Variational Approach to Differential Games" Study No. 52, in Annals of Mathematics, Princeton University Press, Princeton, N. J., 1964.

[B-103] U. Bertele and F. Brioschi, "A New Algorithm for the Solution of the Secondary Optimization Problem in Nonserial Dynamic Programming," J. Math. Anal. Appl. 27, 1969.

[B-104] U. Bertele and F. Brioschi, Parametrization in Nonserial Dynamic Programming, Academic Press, New York, 1969.

[B-105] A. T. Bharucha-Reid, Elements of the Theory of Markov Processes and Their Applications, McGraw-Hill, New York, 1960.

[B-106] D. Blackwell, M. A. Girschick, Theory of Games and Statistical Decisions, Wiley, 1954.

[B-107] G. A. Bliss, Calculus of Variations, University of Chicago Press, Chicago, 1925.

[B-108] V. G. Boltianskii, Mathematical Methods of Optimal Control, Izdat, Nauka, Moscow, 1961; Engl. Transl. by Interscience (Wiley), New York, 1962.

[B-109] V. G. Boltianskii, "Sufficient Conditions for Optimality and the Justification of the Dynamic Programming Method," J. SIAM Control, 4, 1966, pp. 326-361.

[B-110] K. Borch, "Economic Objectives and Decision Problems," IEEE Trans. Sys. Sci. and Cybernet., Sept. 1968, pp. 266-270.

[B-111] T. A. Brown and R. E. Strauch, "Dynamic Programming in Multi-plicative Lattices," JMAA, 12, 1965, pp. 365-370.

[B-112] J. V. Breakwell, J. L. Speyer and E. A. Bryson, "Optimiza-tion and Control of Nonlinear Systems Using the Second Variation," J. SIAM Control, ser. A. Vol. 1, No. 2, March 1963, pp. 193-223.

[B-113] B. M. Budak, E. M. Berkovich, and E. N. Soloveva, "Difference Approximations in Optimal Control Problems," J. SIAM Control, 7, 1969, pp. 18-31.

[B-114] V. N. Burkov and S. E. Lovetskii, "Methods for the Solution of External Problems of Combinatorial Type"(Review) Autom. and Remote Control, no. 11, 1968, pp. 1785-1806.

[B-115] O. Burt and C. Harris, Jr., "Appointment of the U. S. House of Representatives: A Minimum Range, Integer Solution, Allocation Problem," Operations Research, 11, 648-652 (1963).

[B-116] O. Burt, "Optimal Resource Use Over Time with an Application to Ground Water," Management Science, 11, 80-93 (1964).

[B-117] A. G. Butkowsky, A. I. Egonov and K. A. Jurie, "Optimal Control of Distributed Systems (A Survey of Soviet Publications)," J. SIAM Sontrol, 6, 1968, pp. 437-476.

[B-118] A. G. Butkowsky, Distributed Control Processes, Elsevier New York, 1970.

[B-119] T. Butler and A. V. Martin, "On a Method of Courant for Minimizing Functionals," J. Math. & Phys., 41, 1962, pp. 291-299.

[B-120] A. R. Butz, "Convergence with Hilbert's Space Filling Curve," J. Computer Sys. Sci., 3, 1969, pp. 128-146.

[C-1] F. Calogero, Variable Phase Approach to Potential Scattering, Academic Press, New York, 1970.

[C-2] M. D. Canon, C. D. Cullum and E. Polak, Theory of Optimal Control and Mathematical Programming, McGraw-Hill, 1970.

[C-3] C. Caratheodory, Calculus of Variations and Partial Differential Equations of the First Order-I: Partial Differential Equations of the First Order, Holden-Day, San Francisco.

[C-4] S. Cartaino and S. Dreyfus, "Application of Dynamic Programming to the Minimum Time-to-Climb Problem" Aero. Engr. Rev., Vol. 16, No. 6, June 1957, pp. 74-77.

[C-5] A. Charnes and W. W. Cooper, Management Models and Industrial Applications of Linear Programming, Vols. I and II, John Wiley and Sons, New York, 1961.

[C-6] D. Chazan, "Profit Functions and Optimal Control: An Alternate Description of the Control Problem," J. Math. Anal. Appl., 21, 1968, pp. 169-205.

[C-7] H. Chernoff and L. E. Moses, Elementary Decision Theory, John Wiley and Sons, New York, 1959.

[C-8] W. W. Chu, Optimal Adaptive Search, Tech. Rep. 6252-1, Stanford Electronics Laboratories, 1966.

[C-9] C. W. Churchman, R. L. Ackoff, and E. L. Arnoff, Introduction to Operations Research, John Wiley and Sons, New York 1957.

[C-10] M. F. Clement, "Modele Categorique de processes de decisions, theoreme de Bellman, Applications," C. R. Acad. Sci. Paris, 268, 1969, pp, 1-4.

[C-11] L. Collatz, The Numerical Treatment of Differential Equations, Springer, Berlin, 1966.

[C-12] D. C. Collins, "Reduction of Dimensionality in Dynamic Programming via the Method of Diagonal Decomposition," JMAA, 1969.

[C-13] D. C. Collins and A. Lew, "Dimensional Approximation in Dynamic Programming by Structural Decomposition," JMAA, 1970.

[C-14] R. Conti, "Recent Trends in the Theory of Boundary Value Problems for Ordinary Differential Equations," Boll. Unione Mat. Ital., 22(3), 1967, pp. 135-178.

[C-15] R. Courant, D. Hilbert, Methods of Mathematical Physics, Interscience Publishers, 1958.

[C-16] E. D. Conway and E. Hopf, "Hamilton's Theory and Generalized Solutions of the Hamilton-Jacobi Equation," J. Math. Mech., 13, 1964, pp. 939-986.

[C-17] D. R. Cox and H. D. Miller, The Theory of Stochastic Processes, Wiley, 1965.

[C-18] J. Cullum, "Discrete Approximations to Continuous Optimal Control," J. SIAM Control, 7, 1969, pp. 37-49.

[D-1]   G. B. Dantzig, Linear Programming and Extensions, Princeton University Press, New Jersey, 1963.

[D-2]   E. V. Denardo, "Contraction Mappinds in the Theory Underlying Dynamic Programming," SIAM Review, 9, pp. 165-177, April 1967.

[D-3]   E. V. Denardo and L. G. Mitten, "Elements of Sequential Decision Processes," J. Ind. Engr., 18, 1967, pp. 106-112.

[D-4]   C. Derman, Finite State Markovian Decision Processes, Academic Press, 1970.

[D-5]   C. A. Doesoer, "Pontriagin's Maximum Principle and the Principle of Optimality," J. Franklin Inst., 27, 1961, pp. 361-367.

[D-6]   E. W. Dijkstra, "A Note on Two Problems in Connection with Graphs," Numerische Mathematick, 1, 1959.

[D-7]   A. R. Dobell, "Some Characteristic Features of Optimal Control Problems in Economic Theory, IEEE Trans. Auto. Control, AC-14 1969, pp. 39-47.

[D-8] P. Dorato, <u>A Selected Bibliography and Summary of Articles</u>
       <u>on Automatic Control</u>, Res. Rep. PIBMRI-1196-63, Polytechnic
       Institute, Microwave Research Institute, Jan. 1963.

[D-9] S. Dreyfus, "Computational Aspects of Dynamic Programming,"
       <u>Operations Research</u>, 5, 409-415 (1957).

[D-10] S. Dreyfus and M. Freimer, <u>A New Approach to the Duality</u>
       <u>Theory of Mathematical Programming</u>, The RAND Corp. P-2334,
       1961.

[D-11] S. Dreyfus, "The Numerical Solution of Variational Problems,"
       J. Math. Anal. Appl., 5, 1962, pp. 30-45.

[D-12] S. E. Dreyfus, <u>Dynamic Programming and the Calculus of Varia-</u>
       <u>tions</u>, Academic Press, 1965.

[D-13] S. E. Dreyfus, "The Numerical Solution of Nonlinear Optimal
       Control Problems," in <u>Numerical Solutions of Nonlinear Dif-</u>
       <u>ferential Equations</u>, Wiley, New York, 1966, pp. 97-113.

[D-14] S. Dreyfus, <u>An Appraisal of Some Shortest Path Algorithms</u>,
       The RAND Corp., RM-5433-PR, October, 1967.

[D-15] S. E. Dreyfus, "Dynamic Programming and the Hamilton-Jacobi
       Method of Classical Mechanics," <u>J. Optimization Theory and</u>
       <u>Appl.</u>, 2, 1968, pp. 15-26.

[D-16] A. Dvoretsky, J. Kiefer, and J. Wolfowitz, "The Inventory
       Problem: I. Case of Known Distributions of Demand," <u>Econome-</u>
       <u>trica</u>, 20, 450-455, 1952.

[D-17] A. Dvoretzky, J. Kiefer, and J. Wolfowitz, "The Inventory
       Problem: II. Case of Unknown Distributions of Demand,"
       <u>Econometrica</u>, 20, 450-466, 1952.

[E-1] J. E. Eckles, "Optimum Maintenance with Incomplete Infor-
       mation," <u>Opns. Res.</u>, 16, Sept. 1968, pp. 1058-1067.

[E-2] D. O. Ellis, An Abstract Setting for the Notion of Dynamic
       Programming," The RAND Corp, P-783, 1955.

[E-3] S. E. Elmaghraby, "The Concept of 'State' in Discrete Dynamic
       Programming," <u>JMAA</u>, 43, 1973, pp. 642-693.

[E-4] S. E. Elmaghraby and V. Y. Bawle, "Optimization of Batch
       Ordering Under Deterministic Variable Demand," <u>Man. Sci.:</u>
       <u>Theory</u>, 18, pp. 508-517, May 1972.

[E-5] L. E. El'sgol'c, "Variational Problems with a Delayed Argument,"
       <u>Amer. Math. Soc. Trans.</u>, Ser. 2, 16, 1960.

[E-6]   D. Erlenkotter, "Sequencing Expansion Projects," Opns. Res.,
        21, March 1973, pp. 542-553.

[E-7]   A. O. Esogbue and B. R. Marks, "Nonserial Dynamic Programming-
        A Survey," Operational Research Quarterly, 25, 1974, pp. 253-265.

[E-8]   R. V. Evans, "Sales and Restocking Policies in a Single
        Item Inventory System," Man. Sci.: Theory, 14, pp. 463-472,
        March 1968.

[E-9]   R. V. Evans, "Optimality of (s,S) Policies when Set-Up
        Costs Vary," Opsearch, 6, March 1969, pp. 25-46.

[E-10] H. Everett, "Generalized Lagrange Multiplier Method for
        Solving Problems of Optimum Allocation of Resource,"
        Operations Research, 11, 399-417, 1963.

[F-1]   C. H. Falkner, "Jointly Optimal Deterministic Inventory and
        Replacement Policies," Man. Sci.: Theory, 16, May 1970,
        pp. 622-635.

[F-2]   L. T. Fan and C. S. Wang, The Discrete Maximum Principle: A
        Study of Multistage Systems Optimization, John Wiley and Sons,
        New York, 1964.

[F-3]   L. T. Fan et al., "A Sequential Union of the Maximum Principle
        and Dynamic Programming," J. Electron. Control, XVII, 1964,
        pp. 593-600.

[F-4]   A. A. Feldbaum, "Dual Control Theory-I," Automation and Remote
        Control, Vol. 21, No. 9, April 1961, pp. 874-880.

[F-5]   D. Feldman, "Contributions to the Two-Armed Bandit Problem,"
        Ann. Math. Stat., Vol. 33, 1962, pp. 847-856.

[F-6]   W. Feller, "On the Integral Equation of Renewal Theory,"
        Ann. Math. Stat., 21, 1941, pp. 243-267.

[F-7]   W. Fenchel, "On Conjugate Convex Functions," Can. J. Math.,
        1, 1949, pp. 73-77.

[F-8]   F. S. Fine and S. G. Bankoff, "Second Variational Methods
        for Optimization of Discrete Systems," Industrial & Engineer-
        ing Chemistry Fundamentals, 6, May 1967, pp. 293-299.

[F-9]   B. A. Finlayson and L. G. Scriven, "On the Search for Varia-
        tional Principles," Int. J. Heat Mass Transfer, 10, 1967,
        pp. 799-821.

[F-10] S. I. Firstman, B. Gluss, "Optimum Search Routines for Auto-
        matic Fault Location," Ops. Res., Vol. 8, 1960, pp. 512-523.

[F-11] C. E. Flagle, W. H. Huggins, R. H. Roy, Operations Research
        and Systems Engineering, The Johns Hopkins Press, Baltimore,
        1960.

[F-12] I. Fleischer and H. Young, "The Differential Form of the
        Recursive Equation for Partial Optima," J. Math. Anal. Appl.,
        9, 1964, pp. 294-302.

[F-13] I. Fleischer and A. Kooharian, "Optimization and Recursion,"
        J. Soc. Indus. Appl. Math., 12, 1964, pp. 186-188.

[F-14] W. H. Fleming, A Resource Allocation Problem in Continuous
        Form, The RAND Corp., RM-1430, 1955.

[F-15] W. H. Fleming, Discrete Approximations to Some Differential
        Games, The RAND Corp., RM-1526, 1955.

[F-16] M. Florian, M. Klein, "Deterministic Production Planning and
        Concave Costs and Capacity Constraints," Man. Sci.: Theory,
        18, Sept. 1971, pp. 12-20.

[F-17] L. R. Ford, Jr., and D. R. Fulkerson, "Constructing Maximal
        Dynamic Flows from Static Flows," Oper. Res., 5, 1948, pp.
        419-433.

[F-18] G. E. Forsythe, "Today's Computational Methods of Linear
        Algebra," SIAM Rev., 9, 1967, pp. 489-515.

[F-19] R. Fortet, "Properties of Transition Mappings in Dynamic
        Programming," (in French), METRA, 2, 1963, pp. 79-97.

[F-20] B. L. Fox, "Finite State Approximations to Denumerable-State
        Dynamic Programs," J. Math. Analysis and Applications, 34,
        1971, pp. 665-670.

[F-21] B. L. Fox, "Discretizing Dynamic Programs," J. Opt. Theory
        and Appl., 11, 1973, pp. 228-234.

[F-22] J. N. Franklin, Well-posed Stochastic Extensions of Ill
        Posed Linear Problems, Tech. Rep. 135, California Institute
        of Technology, Pasadena, 1969.

[F-23] M. Freimer, M. R. Rao, and M. H. Weingartner, "Note on Muni-
        cipal Bond Coupon Schedules with Limitations on the Number
        of Coupons," Man. Sci., 19, Dec. 1972, pp. 379-380.

[F-24] K. O. Friedrichs, "Ein Verfahren der Variationsrechung,"
        Nachr. Ges. Cottingen, 1929, pp. 13-20.

[F-25] B. L. Fry, Network-Type Management Control Systems Biblio-
        graphy, The RAND Corp., RM-3047-PR, 1963.

[F-26] K. S. Fu, Sequential Methods in Pattern Recognition and
        Machine Learning, Academic Press, New York, 1968.

[F-27] A. T. Fuller, "Bibliography of Optimum Nonlinear Control of
        Deterministic and Stochastic Systems," J. Electron. Control,
        First Series, 13, 1962, pp. 589-611.

[F-28] A. T. Fuller, "Optimization of Some Nonlinear Control Systems
        by Means of Bellman's Equation and Dimensional Analysis,"
        Int. J. Control, 3, 1966, pp. 359-394.

[F-29] A. T. Fuller, "Linear Control of Nonlinear Systems,"
        Int. J. Control. 5, 1967, pp. 197-243.

[G-1]   F. Gantmacher, Theory of Matrices, Chelsea Publishing Co.,
        New York, 1959, Vols. I and II).

[G-2]   J. H. George and W. G. Sutton, "Application of Liapunov
        Theory to Boundary-Value Problems," JMAA to appear.

[G-3]   S. B. Gershwin, "On the Higher Derivatives of Bellman's
        Equation," J. Math. Anal. Appl., 27, 1969.

[G-4]   P. A. Gilmore, "Structuring of Parallel Algorithms," J. Assoc.
        Comp. Mach., 15, 1968, pp. 176-192.

[G-5]   P. C. Gilmore and R. E. Gomory, "Multi-stage Cutting Stock
        Problems of Two or More Dimensions," Opns. Res., 13, 1965,
        pp. 94-120.

[G-6]   I. V. Girsanov, "Certain Relations BEtween the Bellman and
        Krotov Functions for Dynamic Programming Problems," J. SIAM
        Control 7, 1969, pp. 64-67.

[G-7]   C. R. Glassey, "Minimum Change-Over Scheduling of Several
        Products of One Machine," Opns. Res., 16, March 1968,
        pp. 342-352.

[G-8]   B. Gluss, "An Optimum Policy for Detecting a Fault in a Com-
        plex System," Opns. Res., Vol.7, 1959, pp. 468-477.

[G-9]   B. Gluss, "An Optimal Inventory Solution for Some Specific
        Demand Distributions," Naval Research Logistics Quarterly ,
        Vol. 7, 1960, pp. 45-48.

[G-10] B. Gluss, "Lease Squares Fitting of Planes to Surfaces Using
        Dynamic Programming," Comm. Assoc. Computing Machinery, 6,
        1963, pp. 172-175.

[G-11] B. Gluss, "An Alternative Method for Continuous Line Segment
        Curve-Fitting," Inform. Control, 7, 1964, pp. 200-206.

[G-12] B. Gluss, "Generalizations of the Two-Armed Bandit Problem,"
        Ph.D. Thesis, Electrical Engineering Dept., Univ. of Calif.,
        Berkeley, 1965.

[G-13] R. E. Gomory, "The Traveling Salesman Problem," Proc. IBM Sci.
        Comput. Symp. Combinatorial Probl., 1964.

[G-14] R. E. Gomory, "On the Relation Between Integer and Non-Inte-
        ger Solutions to Linear Programs," Proc. Nat. Acad. Sci.,
        53, 260-265.

[G-15] R. H. Gonzalez, "Solution of the Traveling Salesman Problem
        by Dynamic Programming on the Hypercube," Tech. Rep. No.
        18, O.R. Center, M.I.T., 1962.

[G-16] D. Gorman and J. Zaborsky, "Functional Representation of
        Nonlinear Systems, Interpolation and Lagrange Expansion for
        Functionals," ASME Trans., 1966.

[G-17] R. L. Graves and P. Wolfe (eds.), Recent Advances in Mathe-
        matical Programming, McGraw-Hill, New York, 1963.

[G-18] H. J. Greenberg, "Dynamic Programming with Linear Uncertainty,"
        Opns. Res., 16, pp 1100-1114, Nov. 1968.

[G-19] D. E. Greenspan, "Approximate Solution of Initial Value
        Problems for Ordinary Differential Equations by Boundary-
        Value Techniques," J. Math. Phys. Sci., 1, 1967, pp. 261-274.

[G-20] Z. Grihches, "Distributed Lags: A Survey," Econometrica,
        35, 1967, pp. 16-49.

[G-21] R. C. Grinold, "A Generalized Discrete Dynamic Programming
        Model," Man. Sci., 20, pp. 1092-1103, March 1974.

[G-22] J. J. G. Guignabodet, "Marjoration des Errerus de Quantifica-
        tion dans les Calculs de Programmation Dynamique," Compt. Rend.
        255, 1962, pp. 828-830.

[G-23] P. I. Gulyayev, "The Significance of Dynamic Programming and
        Theory of Games for Physiology," (In Russian with English
        Summary), Colloq. Nervous System, 3, LGU Leningrad, 1962,
        pp. 177-189.

[G-24] T. L. Gunckel, "Otpimum Design of Sampled-Data Systems with
        Random Parameters," S.E.L.T.R. No. 2101-2, Stanford Electronics
        Laboratory, Stanford, Calif.

[H-1]   G. Hadley, Linear Programming, Addison-Wesley, Reading,
         Massachusetts, 1962.

[H-2]   G. Hadley and T. Whitin, Analysis of Inventory Systems,
         Prentice-Hall, 1963.

[H-3]   G. Hadley, Nonlinear and Dynamic Programming, Addison-
         Wesley, Reading, Mass., 1964.

[H-4]   Hahn, Susan G., "On the Optimal Cutting of Defective Sheets,"
         Opns. Res., 16, pp. 1100-1114, Dec. 1968.

[H-5]   J. K. Hale and J. P. LaSalle, "Differential Equations: Lin-
         earity vs. Nonlinearity," SIAM Rev., 5, 1963, pp. 249-272.

[H-6]   H. Halkin, "The Principle of Optimal Evolution," in Non-
         linear Differential Equations and Nonlinear Mechanics,
         Academic Press, New York, 1963, pp. 284-302.

[H-7]   R. W. Hamming, Numerical Methods for Scientists and Engineers,
         McGraw-Hill Book Co., Inc., New York, 1962.

[H-8]   H. Hancock, Theory of Maxima and Minima, Dover, N. Y., 1960.

[H-9]   H. Happ, The Theory of Network Diakoptics, Academic Press,
         New York, 1970.

[H-10]   B. K. Harrison, "A Discussion of Some Mathematical Techniques
         Used in Kron's Method of Tearing," J. Soc. Indus. Appl. Math.,
         11, 1963.

[H-11]   N. A. J. Hastings, "Some Notes on Dynamic Programming and
         Replacement," Operat. Res. Quart., 19, pp. 453-464, Dec. 1968.

[H-12]   W. H. Hausman, J. L. Thomas, "Inventory Control with Pro-
         babilistic Demand and Periodic Withdrawals," Man. Sci: Theory,
         18, pp. 265-275, Jan. 1975.

[H-13]   M. Held and R. M. Karp, "A Dynamic Programming Approach to
         Sequencing Problems," J. Soc. Indust. and Appl. Math., 10,1962,
         pp. 196-210.

[H-14]   M. Held, R. M. Karp, and R. Shareshian, "Assembly-Line Balanc-
         ing-Dynamic Programming with Precedence Constraints," Opns.
         Res., 11, 1962, pp. 442-459.

[H-15]   R. Hermann, Differential Geometry and the Calculus of Varia-
         tions, Academic Press, New York, 1968.

[H-16]   H. Hermes and J. P. LaSalle, Functional Analysis and Time
         Optimal Control, New York, 1969.

[H-17] M. Hestenes, Calculus of Variations and Optional Control Theory, Wiley, New York, 1966.

[H-18] T. J. Higgins, "A Resume of the Development and Literature of Nonlinear Control-System Theory," Trans. ASME, April 1957, pp. 445-453.

[H-19] F. B. Hildebrand, Introduction to Numerical Analysis, Mc-Graw-Hill Book Co., Inc., New York, 1956.

[H-20] E. Hille and R. Phillips, Functional Analysis and Semigroups, Amer. Math. Soc. Colloq. Vol. XXXI, 1948.

[H-21] S. Hinatsuka and A. Ichikawa, "Optimal Control of Systems with Transportation Lags," IEEE Trans.Autom. Contr., AC-14, 1969, pp. 237-247.

[H-22] Yu-Chi Ho and R. C. K. Lee, "A Bayesian Approach to Problems in Stochastic Estimation and Control," 1964 JACC, pp. 382-387.

[H-23] R. A. Howard, Dynamic Programming and Markov Processes, John Wiley and Sons, New York, 1960.

[H-24] R. A. Howard, "Information Value Theory," IEEE Trans. Sys. Sci. Cyber., Vol. SSC-2, No. 1, August 1966, pp. 22-26.

[H-25] R. A. Howard, Dynamic Probabilistic Systems, 2, John Wiley and Sons, 1970.

[H-26] A. Hordijk and H. Tijms, "Convergence Results and Approximations for Optimal (s,S) Policies," Man. Sci., 20, July 1974, pp. 1432-1438.

[H-27] D. K. Hughes, "Variational and Optimal Control Problems with Delayed Argument," J. Optim. Theory Appl., 2, 1968, pp. 1-14.

[H-28] W. Hyman, L. Gordon, "Commercial Airline Scheduling Technique," Transportation Research, 2, pp. 23-29, March 1968.

[I-1] T. Ikebe and T. Kato, "Applications of Variational Methods to the Thomas-Fermi Equation," J. Phys. Soc. Japan, 12, 1957, pp. 201-203.

[I-2] E. Ingall, "Optimal Continuous Review Policies for Two Product Inventory Systems with Joint Setup Costs," Man. Sci.: Theory, 15, pp. 278-283, Jan. 1969.

[I-3] P. Ivanescu, "Dynamic Programming with Bivalent Variables," No. 7261, Mat. Vesnik 3(19), 1966, pp. 87-99.

[J-1]   D. H. Jacobson and D. Q. Mayne, Differential Dynamic Program-
        ming, Elsevier, New York, 1970.

[J-2]   D. L. Jaquette, "A Discrete-Time Population-Control Model
        with Setup Cost," Opns. Res., 22, pp. 298-304, March 1974.

[J-3]   P. A. Jensen, "Optimization of Series-Parallel-Series Networks,"
        Opns. Res., 18, pp. 471-482, May, 1970.

[J-4]   P. A. Jensen, "Optimum Network Partitioning," Opns. Res. 19,
        pp. 916-932, May 1971.

[J-5]   C. D. Johnson, "Optimal Control with Chebyshev Minimax
        Performance Index," ASME Trans. J. Basic Eng., June 1967.

[J-6]   E. L. Johnson, "Optimality and Computation of (s,S) Policies
        in the Multi-Item Infinite Horizon Inventory Problem,"
        Man. Sci., 13, Marcy 1967, pp. 475-491.

[J-7]   S. Johnson, "Optimal Two- and Three-Stage Production Schedules
        with Setup Times Included," Nav. Res. Log. Quart., 1, 61-68,
        1954.

[J-8]   P. D. Joseph and T. J.Tou, "On Linear Control Theory," Trans.
        AIEE, Pt. II, Vol. 80, No. 56, pp. 193-196, Sept. 1961.

[J-9]   E. I. Jury and T. Pavlidis, "A Literature Survey of Biocontrol
        Systems," IEEE Trans. on Autom. Control, AC-8, No. 3, July
        1963, pp. 210-217.

[K-1]   S. J. Kahne, "Feasible Control Computations Using Dynamic
        Programming," Proc. Third IFAC Congress, London, June 1966,
        Paper No. 18G, Session 18.

[K-2]   R. Kalaba, "On Nonlinear Differential Equatios, the Maximum
        Operation and Monotone Convergence," J. Math. Mech., 8,
        1959, pp. 519-574.

[K-3]   R. E. Kalman, "On the General Theory of Control Systems,"
        Proc. 1st IFAC Congress, Moscow, USSR, pp. 481-492, Butter-
        worths Press, London, 1960.

[K-4]   R. E. Kalman and R. W. Koepcke, "Optimal Synthesis of Linear
        Sampling Control Systems Using Generalized Performance Indexes,"
        Trans. ASME, Vol. 80, No. 6, Nov. 1958, pp. 1820-1838.

[K-5]   R. E. Kalman and R. S. Bucy, "New Results in Linear Filtering
        and Prediction Theory," Trans. ASME J. Basic Engr., Vol. 83,
        Series D, No. 1, pp. 95-108, March 1961.

[K-6]   R. E. Kalman, "Towards a Theory of Difficulty of Computation
        in Optimal Control," Proc. IBM SCI. Symp. Control Theory Appl.
        1964.

[K-7]   R. E. Kalman, "Algebraic Aspects of the Theory of Dynamical
        Systems," in Differential Equations and Dynamical Systems
        (J. K. Hale and J. P. LaSalle, eds.) Academic Press, New York
        1967, pp. 133-146.

[K-8]   R. E. Kalman, "New Developments in Ssytems Theory Relevant
        to Biology, Systems Theory and Biology," Proc. III Sys. Symp.
        at Case Institute of Technology, Springer, Berlin, 1968.

[K-9]   S. Karlin, Duality in Dynamic Programming, The RAND Corp.
        RM-971, 1952,

[K-10] S. Karlin, "The Structure of Dynamic Programming Models,"
        Naval Res. Logs. Quart., 2, 1955, pp. 285-294.

[K-11] R. M. Karp and M. Held, "Finite-State Processes and Dynamic
        Programming," SIAM J. Appl. Math., 15, 3, 1967.

[K-12] A. Kaufmann, Graphs, Dynamic Programming and Finite Games,
        Academic Press, New York, 1967.

[K-13] A. Kaufmann and R. Cruon, Dynamic Programming: Sequential
        Scientific Management, Academic Press, New York, 1967.

[K-14] W. G. Keckler, R. E. Larson, "Dynamic Programming for the
        Pre-Launch Calculation," Preliminary Report, Contract DA-01-
        021-AMC-90006(y), SRI Project 5188, Stanford Research Insti-
        tute, Menlo Park, Calif., July 1967.

[K-15] W. G. Keckler, "Optimization About a Nominal Trajectory Via
        Dynamic Programming," Engineer's Thesis, Department of
        Electrical Engineering, Stanford, Calif., Sept. 1967.

[K-16] E. B. Keeler, "Projective Metrics and Economic Growth Models,"
        The RAND Corp., RM-6153-PR, Sept. 1969.

[K-17] H. B. Keller, "On the Pointwise Convergence of the Discrete-
        ordinate Method," J. Soc. Indus. Appl. Math., 9, 1960,
        pp. 560-567.

[K-18] J. Kelly, "A New Interpretation of Information Rate," Bell
        System Technical Journal, Vol. 35, 1956, pp. 917-926.

[K-19] J. G. Kemeny and J. L. Snell, Finite Markov Chains, D. Van
        Nostrand, Princeton, New Jersey, 1959.

[K-20] J. D. Kettelle, Jr., "Least-Cost Allocation of Reliability
        Investment," Opns. Res., 10, 1962, pp. 249-265.

[K-21]   M. A. Khvedelidze, "Investigation of the Control Processes
         in Living Organisms and Ways of Developing New Cybernetic
         Systems," (In Russian) Tr. Inst. Kibernet., AN GRUZ SSR, 1,
         1963, pp. 169-190.

[K-22]   D. L. Kleinman, T. Fortmann and M. Athans, "On the Design
         of Linear Systems with Piecewise-Constant Feedback Gains,"
         IEEE Trans. Automat. Control AC-13,   pp. 354-361.

[K-23]   D. E. Knuth, The Art of Computer Programming, Volume I:
         Fundamental Algorithms, Addison-Wesley, 1968.

[K-24]   P. V. Kokotovic and P. Sannuti, "Singular Perturbation Method
         for Reducing the Model Order in Optimal Control Design,"
         IEEE Trans. Automat. Control, AC-13, 1968, pp. 377-383.

[K-25]   R. E. Kopp, "Pontryagin Maximum Principle," in Optimization
         Techniques (G. Leitmann, ed.), Academic Press, New York, 1962.

[K-26]   A. Korsak, Perturbed Optimal Control Problems, Dept. of Math.
         Univ. of Calif., Berkeley, 1966.

[K-27]   A. J. Korsak and R. E. Larson, "A Dynamic Programming Suc-
         cessive Approximations Technique with Convergence Proofs,
         Part I: Description of the Method and Applications, and
         Part II: Convergence Proofs," Automatica, Vol. 6, pp. 245-
         260 (March, 1970).

[K-28]   D. J. Kuck, "Illiac IV Software and Application Programming,"
         IEEE Trans. Computers, C-17, 1968, pp. 758-770.

[K-29]   H. W. Huhn and A. W. Tucker, "Nonlinear Programming," in
         Proc. of the Second Berkeley Symp. on Math. Statistics and
         Probability, 481-490, Univ. of Calif. Press, 1951.

[K-30]   H. W. Kuhn and A. W. Tucker (eds.), "Contributions to the
         Theory of Games II," Annals of Mathematical Studies, 24,
         Princeton University Press, Princeton, New Jersey,1953.

[K-31]   H. J. Kushner, Stochastic Stability and Control, Academic
         Press, New York, 1967.

[K-32]   J. D. R. Kramer, Jr., "On Control of Linear Systems with
         Time Labs," Inform. and Control, 3, 1960, pp. 299-326.

[K-33]   N. N. Krasovskii, "On the Analytical Construction of an
         Optimal Control in Systems with Time Lags," J. Appl. Math.
         Mech., 26, 1962, pp. 50-67.

[K-34]   H. G. L. Karuse, Astrorelativity, NASA Tech. Rep. TR R-188,
         Jan. 1964.

[K-35]   P. Krolak and L. Cooper, "An Extension of Fibonaccian Search to Several Variables," Comm ACM, 6, 1963, pp. 369-641.

[K-36]   E. Kreindler, "Reciprocal Optimal Control Problems," J. Math. Appl., 14, 1966, pp. 141-152.

[K-37]   G. Kron, A Set of Principles to Interconnect the Solution of Physical Systems," J. Appl. Phys., 24, 1953, pp. 965-980.

[K-38]   G. Kron, "Tearing and Interconnecting as a Form of Transformation," Quart. Appl. Math., 13, 155, pp. 147-159.

[K-39]   V. F. Krotov, "Methods of Solving Variational Problems on the Basis of the Sufficient Conditions Governing the Absolute Minimum," Automat. Telemekh.23, 1962, pp. 1473-1484.

[K-40]   E. Kruger-Thiemer and R. R. Levine, "The Solution of Pharmacological Problems with Computers," Drug. Res., 18, 1968, pp. 1575-1579.

[L-1]   R. B. Laning, Studies Basic to the Consideration of Artificial Heart Research and Development Program, Final Rep. PB 169-831, US Dept. of Commerce, National Bureau of Standards, Inst. for Applied Technology, Washington, D.C., 1961.

[L-2]   P. A. Larkin, "The Possible Shape of Things to Come," SIAM Rev., 11, 1969, pp. 1-6.

[L-3]   R. E. Larson, F. J. Rees and J. P. Nichols, "Dynamic Programming for Expansion Planning of Nuclear Power Generation Systems," American Nuclear Society Transactions, Vol. 13, No. 1, pp. 37-38 (June-July, 1970).

[L-4]   R. E. Larson, Dynamic Programming with Continuous Independent Variable, Stanford Electronics Laboratory, TR 6302-6, Stanford, Calif., April 1964.

[L-5]   R. E. Larson, "Dynamic Programming with Reduced Computational Requirements," IEEE Trans. on Automatic Control, Vol. AC-10 No. 2, April 1965, pp. 135-143.

[L-6]   R. E. Larson, "An Approach to Reducing the High Speed Memory Requirement of Dynamic Programming," J. Math. Anal. and Appl., Vol. 11, Nos. 1-3, July 1965, pp. 519-537.

[L-7]   R. E. Larson and J. Peschon, "A Dynamic Programming Approach to Trajectory Estimation," IEEE Trans. Auto. Contr., Vol. AC-10, No. 3, July 1966, pp 537-540.

[L-8]   R. E. Larson, "A Survey of Dynamic Programming Computational Procedures," IEEE Transactions on Automatic Control, Vol. AC-12, No. 6, pp 767-774 (December, 1967).

[L-9]   R. E. Larson and W. G. Keckler, "Applications of Dynamic
         Programming to the Control of Water Resource Systems,"
         Automatica, Vol. 5, No. 1, pp. 15-26 (January, 1969).

[L-10]   R. E. Larson and P. J. Wong, "Optimization of Natural-Gas
         Pipeline Systems Via Dynamic Programming," IEEE Transactions
         on Automatic Control, Vol. AC-13, No. 5, pp. 475-481
         (October, 1968).

[L-11]   R. E. Larson, R. M. Dressler, and R. S. Ratner, "Precomputa-
         tion of the Weighting Matrix in an Extended Kalman Filter,"
         1967 Joint Automatic Control Conference, University of
         Pennsylvania, Philadelphia, Pennsylvania (June, 1967).

[L-12]   R. E. Larson, L. Meier, and A. J. Tether, "Dynamic Program-
         ming for Stochastic Control of Discrete Systems," IEEE Trans-
         actions on Automatic Control, Vol. AC-16, No. 6 (Dec., 1971).

[L-13]   R. E. Larson and E. Tse, "Parallel Processing Algorithms for
         the Optimal Control of Nonlinear Dynamic Systems," IEEE
         Transactions on Computers, Vol. C-22, No. 8 (August, 1973).

[L-14]   R. E. Larson, State Increment Dynamic Programming, Elsevier,
         New York, 1968.

[L-15]   R. E. Larson, Y. Bar-Shalom, and M. A. Grossberg, "Applica-
         tion of Stochastic Control Theory to Resource Allocation
         Under Uncertainty," IEEE Transactions on Automatic Control,
         Vol. 19, No. 1 (February, 1974).

[L-16]   R. E. Larson and W. G. Keckler, "Optimum Adaptive Control
         in an Unknown Environment," IEEE Trans. Auto. Control,
         Vol. AC-13, No. 4, August 1968, pp. 438-439.

[L-17]   J. P. LaSalle, "Stability and Control," J. SIAM Control, 1,
         1962, pp. 3-15.

[L-18]   R. Lattes and J. L. Lions, The Method of Quasi-reversibility:
         Applications to Partial Differential Equations, Elsevier,
         New York, 1969.

[L-19]   E. L. Lawler, "Partitioning Methods for Discrete Optimization
         Problems," in Recent Mathematical Advances in Operations
         Research, Univ. of Michigan, Ann Arbor, 1964.

[L-20]   P. D. Lax, "Hyperbolic Systems of Conservation Laws II,"
         Comm. Pure Appl. Math., 10, 1957, pp. 537-566.

[L-21]   E. S. Lee, "Dynamic Programming, Quasilinearization and
         Dimensionality Difficulty," J. Math. Anal. Appl., 27, 1968,
         pp. 303-322.

[L-22] R. C. K. Lee, Optimal Estimation, Identification and Control,
        MIT Press, 1964.

[L-23] A. Lew, "Approximation Techniques in Discrete Dynamic Program-
        ming," TR70-10, Univ. Southern Calif., Dept. of Elec. Eng.,
        January 1970.

[L-24] A. Lew, "Reduction of Dimensionality by Approximation Techni-
        ques: Diffusion Processes," J. Math. Anal. Appl., 1972.

[L-25] D. V. Lindley, Probability and Statistics, Vol. 2, Cambridge
        Univ. Press, pp. 59-72, 1964.

[L-26] L. D. Liozner, "Regeneration of Lost Organs," Izdatel's
        stvu Akad. Nauk SSR, Moscow, 1962, pp. 1-141; Engl. Trans.,
        FTD-TT 63-576, Div. Foreign Technology, Air Force Systems
        Command, Wright-Patterson AFB, Sept. 1963.

[L-27] G. G. Lorentz, "Metric Entropy and Approximation," Bull.
        Amer. Math. Soc., 72, 1966, pp. 903-937.

[L-28] A. G. Lubowe, "Optimal Functional Approximation Using
        Dynamic Programming," AIAA J., 2, 1964, pp. 376-377.

[L-29] R. D. Luce and H. Raiffa, Games and Decisions: Introduction
        and Critical Survey, John Wiley and Sons, New York 1957.

[L-30] D. L. Lukes, "Optimal Regulation of Nonlinear Dynamical
        Systems," J. SIAM Control, 7, 1969, pp. 75-100.

[L-31] H. Luss and Z. Kander, "Inspection Policies when Duration
        of Checking is Non-Negligible," Operat. Res. Quart., 25,
        June 1974, pp. 299-309.

[M-1]   S. MacLane, "Hamiltonian Mechanics and Geometry," Amer. Math.
        Monthly, 77, 1970, pp. 570-585.

[M-2]   J. MacQueen, "A Test for Suboptimal Actions in Markovian
        Decision Problems," Opns. Res., 15, pp. 559-561, May 1967.

[M-3]   R. B. Maffei, "Planning Advertising Expenditures by Dynamic
        Programming Methods,"Management Technology,1,1960,pp. 94-100.

[M-4]   M. Magazine, Optimal Policies for Queueing Systems with
        Periodic Review, Ph.D. Thesis, Univ. of Fla, 1969.

[M-5]   O. Mangasarian, Nonlinear Programming, Mc-Graw-Hill, 1969.

[M-6]   D. Mangeron, "The Bellman Equations of Dynamic Programming
        Concerning a New Class of Boundary Value Problems with Total
        Derivatives," J. Math Anal. Appl., 8, 1964, pp. 141-146.

[M-7]   D. Mangeron, "Sur une classe d'equations fonctionnelles
        attaches aux problems extremaux correspondant aux systemes
        polyvibrants," C. R. Acad. Sci. Parris, 266, 1968, pp. 1121-
        1124.

[M-8]   A. S. Manne, "Linear Programming and Sequential Decisions,"
        Management Science, 6, 1960, pp. 259-267.

[M-9]   P. Masse, Les reserves et la regulation de l'avenir dans la
        vie economique; Vol. I: Avenir determine; Vol II; Avenir
        aleataire, Hermann et Cie, Paris, 1946.

[M-10] S. R. McReynolds and A. E. Bryson, "A Successive Sweep
        Method for Solving Optimal Programming Problems," Joint
        Auto. Control Conf., Troy, N. Y. 1965.

[M-11] S. R. McReynolds, "The Successive Sweep Method and Dynamic
        Programming," J. Math. Anal. Appl., 18, 1967, pp. 355-364.

[M-12] L. Meier, "Combined Optimum Control Estimation," paper
        presented at 1965 Allerton Conference, University of Illinois,
        pp. 109-120.

[M-13] V. W. Merriam, "A Class of Optimum Control Systems," J.
        Franklin Inst., 267, 1959, pp. 367-281

[M-14] C. F. Meyer and R. J. Newett, "Dynamic Programming for
        Feedlot Optimization," Man. Sci.: Application, 16, pp. B-410
        Feb. 6, 1970.

[M-15] A. Miele (ed.), Theory of Optimum Aerodynamic Shapes,
        Academic Press, New York, 1965.

[M-16] H. Mine and S. Osaki, Markovian Decision Problems, Elsevier,
        New York 1970.

[M-17] G. J. Minty, "Monotone Networks," Proc. Roy Soc. London,
        Ser. A, 257, 1960, pp. 194-212.

[M-18] L. G. Mitten, "An Analytic Solution to the Least Cost Testing
        Sequence Problem," J. of Ind. Eng., 11, 17, 1960.

[M-19] L. G. Mitten and G. L. Menhauser, "Multistage Optimization,"
        Chemical Engineering Progress, 59, 1963, pp. 52-60.

[M-20] L. G. Mitten and G. L. Nemhauser, "Optimization of Multi-
        stage Separation Processes by Dynamic Programming," Can. J.
        of Chemical Eng., 41, 1963, pp 187-194.

[M-21] L. G. Mitten, "Composition Principles for Syntehsis of
        Optimal Multi-stage Processes," Opns. Res., 12, 1964, pp.
        610-619.

[M-22] L. G. Mitten, "Preference Order Dynamic Programming,"
        Man. Sci., 21, Sept. 1974, pp. 43-46.

[M-23] N. N. Moiseev, "Methods of Dynamic Programming in the Theory
        of Optimal Controls (IV)," USSR Computa. Math. Phys., 5,
        1965, pp. 58-75.

[M-24] J. Monavek, "A Note on Minimal Path Problems," JMAA, 1970.

[M-25] B. Mond and M. A. Hanson, "Duality for Variational Problems"
        J. Math. Anal. Appl., 18, 1967, pp. 355-364.

[M-26] M. Morse, Calculus of Variations in the Large, American
        Mathematical Society, 1935.

[M-27] J. J. Moreau, "Convexity and Duality," Functional Analysis
        and Optimization, Academic Press, New York, 1966, pp. 145-169.

[M-28] J. J. Moreau, "Fonctions convexes duales et points proximaux
        dans un espace Hilbertien," C. R. Acad. Sci., Paris 255,
        1962, pp. 2897-2899.

[M-29] J. J. Moreau, "Analyse Fonctionelle, Inf.-convolution des
        fonctions numeriques sur un espace vectoriel," C. R. Acad.
        Sci., 256, pp. 5047-5049.

[M-30] J. J. Moreau, "Proprietes des applications 'prox'," C. R.
        Acad. Sci., Paris, 256, 1963, pp. 1069-1071.

[M-31] T. L. Morin, A. O. Esogbue, "The Imbedded State Space
        Approach to Reducing Dimensionality in Dynamic Programs of
        Higher Dimensions," J. Math Analysis and Appl., 148, 1974,
        pp. 801-811.

[M-32] R. Morton, "On the Dynamic Programming Approach to Pontriagin's
        Maximum Principles," J. Appl. Prob., 5, 1968, pp. 679-692.

[M-33] T. E. Morton, "On the Asymptotic Convergence Rage of Cost
        Differences for Markovian Decision Processes," Opns. Res.,
        19, Jan. 1971, pp. 244-248.

[M-34] J. Moser, "A New Technique for the Construction of Solutions
        of Nonlinear Differential Equations," Proc. Nat. Acad. Sci.
        USA, 47, 1961, pp. 1824-1831.

[N-1]   G. L. Nemhauser, "Applications of Dynamic Programming in the
        Process Industries," Am. Inst. Ind. Eng. Proceedings., 1963,
        pp. 279-292.

[N-2]   G. L. Nemhauser, "Decomposition of Linear Programs by Dynamic
        Programming," Naval Research Logistics Quart, 11, 1964,
        pp. 191-196.

[N-3]   G. L. Nemhauser and Z. Ullman, "Discrete Dynamic Programming
        and Capital Allocation," Manag. Sci., 15, 1969, pp. 494-505.

[N-4]   G. L. Nemhauser, Introduction to Dynamic Programming, John
        Wiley and Sons, 1966.

[N-5]   G. L. Nemhauser, "Scheduling Local and Express Service,"
        Trans. Sci., 3, May 1969, pp. 164-175.

[N-6]   G. L. Nemahuser and P. LuYu, "A Problem in Bulk Service
        Scheduling," Opns. Res., 20, July 1972, pp. 813-819.

[N-7]   E. Noether, "Invariant Variationsprobleme," Nachr. Ges.,
        Gottingen, 1918, pp. 235-237.

[O-1]   A. Odoni, "On Finding the Maximal Gain for Markovian Decision
        Processes," Opns. Res., 17, Sept. 1969, pp. 857-860.

[O-2]   C. Olech, "Existence Theorems for Optimal Problems with
        Vector-Valued Cost Functions," Trans. Amer. Math. Soc., 136
        1969, pp. 159-180.

[O-3]   H. Osborn, "Euler Equations and Characteristics," Chpt. 7
        in Dynamic Programming of Continuous Processes, (R. Bellman,
        ed.) the RAND Corp, R-271, 1954.

[O-4]   A. Ostrowsky, "On Two Problems in Abstract Algebra Connected
        with Horner's Rule," Studies in Mathematics and Mechanics
        Presented to Richard von Mises, Academic Press, New York,
        1954, pp. 40-48.

[P-1]   V. G. Pavlov and V. P. Cheprasov, "Constructing Certain
        Invariant Solutions of Bellman's Equation," Automat. Remot.
        Control., January 1968, pp. 31-36.

[P-2]   J. D. Pearson, "Reciprocity and Duality in Control Programming
        Problems," J. Math. Anal. Appl., 1965, pp. 385-408.

[P-3]   V. Pereyra, "On Improving an Approximate Solution of A
        Functional Equation by Deferred Corrections," Numer. Mat. 8,
        1966, pp. 376-391.

[P-4]   J. Peschon, Disciplines and Techniques of Systems Control,
        Blaisdell Publishing Co., New York, 1965.

[P-5]   J. Peschon and R. E. Larson, "Analysis of an Intercept System,"
        SRI, Menlo Park, Calif, Final Report to NIKE-X Project Office,
        Redstone Arsenal, Alabama, Contract DA-01-021-AMC-900005(y),
        SRI Project 5188-7, December 1965.

[P-6]   E. R. Peters, "A Dynamic Programming Model for the Expansion
        of Electric Power Systems," Man. Sci., 10, Dec. 1973, pp.
        656-664.

[P-7]   R. Petrovic, "Optimization of Resource Allocation in Project
        Planning," Opns. Res., 16, May 1968, pp. 559-568.

[P-8]   G. Polya and G. Szego, Augfaben und Lehrsatze aus der Analysis,
        Dover, New York, 1945.

[P-9]   M. Pollack and W. Wiebenson, "Solution of the Shortest Route
        Problem - A Review," Operations Research, 8, 1960, pp.224-230.

[P-10] G. C. Pomraming, "Reciprocal and Canonical Forms of Variational
        Problems Involving Linear Operators," J. Math. Phys., XLVII,
        1968.

[P-11] L. S. Pontryagin et al., The Mathematical Theory of Optimal
        Processes, John Wiley & Sons, Inc., New York, 1962.

[P-12] E. L. Porteus, "Some Bounds for Discounted Sequential Deci-
        sion Progress," Man. Sci., 1971, pp. 7-11.

[P-13] F. Prochan and T. A. Bray, "Optimal Redundancy Under Multiple
        Constraints," Opns. Res., 13, 1965, pp. 800-814.

[P-14] B. A. Powell, "Optimal Elevator Banking Under Heavy Up-
        Traffic," Trans. Sci., 5, May 1971, pp. 109-121.

[P-15] M. J. Protter, "Difference Methods and Soft Solutions,"
        Nonlinear Partial Differential Equations, Academic Press,
        New York, 1963, pp. 161-170.

[P-16] E. B. Pyle, III, B. Douglas, G. W. Ebright, W. J. Westlake,
        A. D. Bender, "Scientific Manpower Allocation to New Drug
        Screening Programs," Man. Sci., 19, Aug. 1973, pp. 1433-1443.

[R-1]   R. Radner, "Paths of Economic Growth that are Optimal with
        Regard Only to Final States: A 'Turnpike Theorem',"
        Rev. Econ. Stud., 28, 1961, pp. 98-104.

[R-2]   L. B. Rall, "On Complementary Variational Principles," J.
        Math. Anal. Appl., 14, 1966, pp. 174-184.

[R-3]   V. R. Rao, J. L. Thomas, "Dynamic Models for Sales Promotion
        Policies," Operat. Res. Quart., 24, Sept. 1973, pp. 403-417.

[R-4]   R. S. Ratner, "Performance-Adaptive Renewal Policies for the
        Operation of LInear Systems Subject to Failure," Ph.D.
        Thesis, Dept. of Elec. Engr., Stanford Univ., Calif, June 1968.

[R-5]   H. E. Rauch, F. Tung, and C. G. Streibel, "Maximum Likelihood
         Estimates of Linear Dynamic Systems," AIAA J., Vol. 3,
         No. 8, Aug. 1965, pp. 1445-1450.

[R-6]   F. J. Rees and R. E. Larson, "Computer-Aided Dispatching
         and Operations Planning for an Electric Utility with
         Multiple Types of Generation," IEEE Transactions on Power
         Apparatus and Systems, Vol. PAS-90, No. 2, (March-April,
         1971).

[R-7]   D. W. Richardson, E. P. Hechtman, R. E. Larson, "Control
         Data Investigation for Optimization of Fuel on Supersonic
         Transport Vehicles," Phase II Report, Contract AF 33(657)-
         8822, Project 9056, Hughes Aircraft Co., Culver City, Calif.,
         June 1963.

[R-8]   R. D. Richtmyer, Difference Methods for Initial Value Problems,
         Interscience Publishers, Inc., New York, 1957.

[R-9]   V. Riley and S. I. Gass, Linear Programming and Associated
         Techniques, The Johns Hopkins University Press, Baltimore
         1958.

[R-10]   S. M. Roberts, Dynamic Programming in Chemical Engineering
         and Process Control, Academic Press, New York, 1964.

[R-11]   R. T. Rockafeller, "Conjugates and Legendre Transforms of
         Convex Functions," Can. J. Mat., 19, 1967, ppl 200-205.

[R-12]   R. T. Rockafeller, "Duality and Stability in Extremum Problems
         Involving Convex Functions," Pacific J. Math., 21, 1967,
         pp. 200-205.

[R-13]   R. T. Rockafeller, Convexity and Duality, Princeton Univ.
         Press, Princeton, N. J. 1968.

[R-14]   R. J. Roman, "Mine-Mill Production Scheduling by Dynamic
         Programming," Operat. Res. Quart, 22, Dec. 1971, pp. 319-328.

[R-15]   S. H. Ross, Applied Probability Models with Optimization
         Applications, Holden-Day, San Francisco 1970.

[R-16]   M. Rosseel, "Comments on a Paper by Romesh Saigal: 'A
         Constrained Shortest Route Problem',", Opns. Res., 16, 1968,
         pp. 1232-1234.

[R-17]   C. J. Rose, "Dynamic Programming Processes within Dynamic
         Programming Processes, JMAA.

[R-18]   P. Rosenblook, "The Method of Steepest Descent," Proc. Symps.
         Appl. Math., VI, 1956, pp. 127-176.

[R-19] R. Roth, "The Unscrambling of Data Studies in Segmental Dif-
        ferential Approximation," J. Math. Anal. Appl., 14, 1966,
        pp. 5-22.

[R-20] R. Roth, "An Application of Algebraic Topology: Kron's
        Method of Tearing,"Quart. Appl. Math., 17, 1959, pp. 1-14.

[R-21] M. Rothstein, "An Airline Overbooking Model," Trans. Sci.,
        5, May 1971, pp. 180-192.

[R-22] L. I. Rozonoer, "The Maximum Principle of L. S. Pontryagin
        in Optimal System Theory,"Automat. Remote Control, 20, 1960,
        pp. 1288-1302; 1517-1532.

[R-23] L. I. Rozonoer, "Variational Approach to the Problem of
        Invariance of Automatic Control Systems-I," Avtomat. Telemekh.
        24, 1963, pp. 774-756.

[R-24] D. P. Rutenberg, "Design Commonality to Reduce Multi-Item
        Inventory: Optimal Depth of a Product Line," Opns. Res., 19
        March 1971, pp. 491-509.

[S-1]   T. L. Saaty, Mathematical Methods of Operations Research,
        McGraw-Hill, New York, 1959.

[S-2]   P. Sannuto and P. V. Kokotovis, "Near Optimum Design of
        Linear Systems by a Singular Perturbation Method," IEEE Trans.
        Autom. Control, AC-14, 1969, pp. 15-21.

[S-3]   L. J. Savage, The Foundations of Statistics, Wiley, 1954.

[S-4]   V. K. Savley, Integration of Equations of Parabolic Type by
        the Method of Nets, Trans. by G. S. Tee (Macmillan Co.,
        New York, 1964.

[S-5]   H. E. Scarf, D. M. Gilford and M. W. Shelly (eds.), Multi-
        Stage Inventory Models and Techniques, Stanford University
        Press, Stanford, Calif, 1963.

[S-6]   A. Schronhage, "Multiplication grosser Zahlen," Computing
        (Arch. Elektron, Rechnen), 1, 1966, pp. 182-196.

[S-7]   P. J. Schweitzer, "Multiple Policy Improvements in Undis-
        counted Markov Renewal Programming," Opns. Res., 19, May 1971,
        pp. 784-793.

[S-8]   J. F. Shapiro, Shortest Route Methods for Finite State Space,
        Deterministic Dynamic Programming Problems, Tech. Rep. Oper-
        ations Research Center, MIT, Cambridge, Mass. 1967.

[S-9]   J. F. Shapiro, "Dynamic Programming Algorithms for the
        Integer Programming Problem-1: The Integer Programming
        Problem Viewed as a Knapsack Type Problem," Opns. Res., 16,
        Jan. 1968, pp. 103-121.

[S-10] J. F. Shapiro, "Turnpike Theorems for Integer Programming
        Problems," Opns. Res., 18, May 1970, pp. 432-440.

[S-11] M. W. Shelley, II, and G. L. Bryan (eds.), Human Judgements
        and Optimality, Wiley, New York, 1964.

[S-12] L. B. Slobodkin, "The Stragety of Evolution," Amer. Sci., 52
        1964, pp. 342-357.

[S-13] F. J. Smith, "An Algorithm for Summing Orthogonal Polynomial
        Series and Their Derivatives with Applications to Curve-
        Fitting and Interpolation," Math. Comp., 19, 1965, pp. 33-36.

[S-14] M. Sobral, "Sensitivity in Optimal Control Systems,"
        Proc. IEEE. 45, 1968, pp. 1644 -1752.

[S-15] W. R. Spillers and N. Hickerson, "Optimal Elimination for
        Sparse Symmetric Systems as a Graph Problem," Quart. Appl.
        Math., 26, 1968, pp. 425-432.

[S-16] V. P. Sreedharan, H. H. Wein, "A Stochastic, Multi-stage
        Investment Model," SIAM J. Appl. Math., 15, Marcy 1967,
        pp. 347-358.

[S-17] D. V. Stewart, "Partitioning and Tearing Systems of Equations,"
        J. Soc. Indust. Appl. Math., Number Anal. 2, 1965, pp. 345-365

[S-18] A. W. J. Stoddart, "Estimation of Optimality for Multidimen-
        sional Control Systems," J. Optimization Theory and Appl.,
        3, 1969, pp. 385-391.

[S-19] H. Stone, "Approximation of Curves by Line Segments," Math
        of Comp., 15, 1961, pp. 40-47.

[S-20] A. R. Stubberud and J. M. Swiger, "Minimum Energy Control of
        a Linear Plant with Magnitude Constraint on the Control
        Input Signals," 1965, JACC, pp. 308-406.

[S-21] J. Sugie, "An Extension of Fibonaccian Searching to the
        Multidimensional Cases," IEEE Trans. Automat. Control, AC-9,
        Jan. 1964.

[S-22] R. Sussman, "Otpimal Control of Systems with Stochastic
        Disturbances," Technical Report Series 63, No. 20, Electronics
        Laboratory, University of California, Berkeley, 1963.

[T-1]   A. Tchamram, "A New Derivation of the Maximum Principle,"
        J. Math. Anal. Appl., 25, 1969, pp. 350-361.

[T-2]   L. Tisza, "The Conceptual Structure of Physics," Rev. Mod.
        Phys., 35, 1963, pp. 151-185.

[T-3]   J. Todd, "Motivation for Working in Numerical Analysis,"
        Comm. Pure Appl. Math, 8, 1955, pp. 97-116.

[T-4]   J. Todd, "The Problem of Error in Digital Computation,"
        Error Digital Computat, 1, 1965, pp. 3-41.

[T-5]   J. T. Tou, Optimum Design of Digital Control Systems,
        Academic Press, N. Y. 1963.

[T-6]   R. Tomovic, and G. Boni, "An Adaptive Artificial Hand,"
        IEEE Trans. on Autom. Control, AC-7, No. 3, April 1962,
        pp. 3-10.

[T-7]   R. Tomovic and L. Radanopvic, "Homeostatic Control of Dynamic
        Systems," Proc. First Int. Symp. Optimization Adap. Control,
        April 1962, pp. 57-67.

[V-1]   P. J. M. Van den Bogaard, A. M. Luque and C. Van de Panne,
        "A Study of the Implications of Alternative Layers in Quadra-
        tic Dynamic Programming," Rev. Fr. de Recherche Operationelle
        6, 1962, pp. 163-183.

[V-2]   A. F. Veinott, Jr., "On Finding Optimal Policies in Discrete
        Dynamic Programming with No Discounting," Ann. Math'l. Stat.,
        37, 1966, pp. 1284-1295.

[V-3]   A. F. Veinott, Jr., "Minimum Concave-Cost Solution of Leontief
        Substitution Models of Multi-Facility Inventory Systems,"
        Opns. Res., 17, pp. 262-291, March 1969.

[V-4]   T. L. Vincent and J. D. Mason, "Disconnected Optimal Trajec-
        tories," JOTA, 3, 1969, pp. 263-281.

[V-5]   T. Von Karman, "The Engineer Grapples with Nonlinear Problems,"
        Bull. Amer. Math. Soc., 46, 1940, pp. 615-683.

[V-6]   J. Von Neumann and O. Morgensterm, Theory of Games and Econo-
        mic Behavior, Princeton Univ. Press, Princeton, N. J. 1944.

[W-1]   A. Wald, Sequential Analysis, Wiley, 1947.

[W-2]   A. Wald, Statistical Decision Functions, John Wiley & Sons,
        1950.

[W-3]   P. K. C. Wang, "Optimal Control of Distributed Control
        Systems with Time Delays," IEEE Trans. Automat. Control, AC-9,
        1964, pp. 13-22.

[W-4]   P. K. C. Wang, "Control of Distributed Parameter Systems," in
        Advances in Control Systems, Vol. 1, Academic Press, New
        York, 1964, pp. 75-172.

[W-5]   P. K. C. Wang and F. Tung, "Optimal Control of Distributed
        Parameter Systems," J. Basic Eng. Trans. ASME, SER D, 86,
        1964, pp. 67-79.

[W-6]   K. E. F. Watt, "Dynamic Programming, Look Ahead Programming
        and the Strategy of Insect Pest Control," Can. Entomol.95,
        1963, pp. 525-536.

[W-7]   K. E. F. Watt, "Computers and the Evaluation of Resource
        Management Strategies," Amer. Sci., 62, 1964, pp. 408-419.

[W-8]   H. M. Weingartner, "Capital Budgeting of Interrelated Problems:
        Survey and Synthesis," Man. Sci., 12, 1966, pp. 485-516.

[W-9]   H. M. Weingartner, "Municipal Bond Coupon Schedules with
        Limitations on the Number of Coupons," Man. Sci., 19, Dec.
        1972, pp. 369-378.

[W-10] J. H. Westcott, J. J. Florentin and J. D. Pearson, "Approx-
        imation Methods in Optimal Adaptive Control," Proc. IFAC
        Sec. Congr. Basel, 1963.

[W-11] D. M. White, "Dynamic Programming and Probabilistic Con-
        straints," Opns. Res., 22, May 1974, pp. 654-664.

[W-12] P. Whittle, "The Deterministic Stochastic Transition in
        Control Processes and the Use of Maximum and Integral Trans-
        forms," in Proc. Fifth Berkeley Symp. Math. Stat. Prob.,
        Vol. II, Univ. of California Press, Berkeley, 1967.

[W-13] P. Whittle, "A View of Stochastic Control Theory," J. Royal
        Stat. Soc.,132, 1969, pp. 320-335.

[W-14] C. Wilkinson, S. K. Gupta, "Allocating Promotional Effort
        to Competing Activities: A Dynamic Programming Approach,"
        Proc. of Fifth International Conf. on Opns. Res., June 1969.
        pp. 419-432.

[W-15] D. M. Wiberg, "Feedback Control of Linear Distributed Systems,
        Trans. ASME, 89, 1967, pp. 379-384.

[W-16] N. Wiener, "Oroblems of Sensory Prosthesis," Bull. Amer.
        Math. Soc., 57, No. 1, Part 1. Jan. 1951.

[W-17] D. J. Wilde, Optimum Seeking Methods, Prentice-Hall,
        Englewood Cliffs, New Jersey, 1964.

[W-18] D. J. Wilde and C. S. Beightler, Foundations of Optimization,
        Prentice Hall, New York, 1967.

[W-19] D. M. G. Wishart, "A Survey on Control Theory," J. Royal
        Stat. Soc., 132, 1969, pp. 293.

[W-20] L. A. Wolsey, "Generalized Dynamic Programming Methods in
        Integer Programming," Math. Prog., 4, April 1973, pp. 222-232.

[W-21] P. J. Wong, "Dynamic Programming Using Shift Vectors," Ph.D.
        Thesis Dept. of Elec. Engr., Stanford University, Stanford,
        Calif, Sept. 1967.

[W-22] P. J. Wong and D. G. Luenberger, "Reducing the Memory
        Requirements of Dynamic Programming," Oper. Res., 16, 1968,
        pp. 1115-1125.

[W-23] P. J. Wong, "A New Decomposition Procedure for Dynamic Pro-
        gramming," Oper. Res., 18, 1970, pp. 119-131.

[W-24] P. J. Wong, "An Approach to Reducing the Computing Time for
        Dynamic Programming," Opns. Res., 19, Jan. 1970, pp. 181-185.

[W-25] W. M. Wonham, "Random Differential Equations in Control Theory"
        in Probabilistic Methods in Applied Mathematics, Vol. 2,
        Academic Press, 1970, pp. 131-212.

[W-26] R. D. Woodman, "Replacement Policies for Components that
        Deteriorate," Operat. Res. Quart., 18, Sept. 1967, pp. 267-281.

[W-27] A. Wouk, "Approximation and Allocation," J. Math. Anal.
        Appl., 8, 1964, pp. 135-143.

[W-28] A. Wouk, "An Extension of the Caratheodory-Kalman Variational
        Method," J. Optimization Theory Appl. 3, 1968, pp. 2-34.

[Y-1]   S. Yakowitz, Mathematics of Adaptive Control Processes,
        American Elsevier, Co., New York, 1969.

[Y-2]   Y. B. Yasinsky and S. Kaplan, "On the Use of Dual Variational
        Principles for the Estimation of Error in Approximate Solu-
        tions of Diffusion Problems," Nuc. Sci. Eng., 31, 1968,
        pp. 80-90.

[Y-3]   D. R. Young, "Scheduling a Fixed-Schedule, Common Carrier
        Passenger Transportation System," Trans. Sci., 4, Aug. 1970.
        pp. 243-269.

[Y-4]   L. C. Young, Lectures on the Calculus of Variations and
         Optimal Control Theory, Saunders, Philadelphia 1968.

[Z-1]   J. Zaborsky and D. Gorman, "Control by Functional Lagrange
         Expansions," IFAC, 1966.

[Z-2]   S. Zacks, "A Two-Echelon Multi-Station Inventory Model for
         Navy Applications," Naval Research Logistics Quarterly,
         17, March 1970, pp. 79-85.

[Z-3]   L. Zadeh, "What is Optimal?" IEEE Trans. Inform. Theory,
         IT-4, 1958, p. 3.

[Z-4]   L. Zadeh, C. A. Doesoer, Linear System Theory: The State
         Space Approach, McGraw-Hill, 1963.